宅在台湾 IV

HOUSES IN TAIWAN

INTERPRETING TAIWANESE SPACE IMPRESSION

解　读　台　式　空　间　印　象

中国林业出版社
China Forestry Publishing House

图书在版编目（CIP）数据

宅在台湾. Ⅳ / 深圳视界文化传播有限公司编. --
北京 : 中国林业出版社, 2017.7
ISBN 978-7-5038-9115-1

Ⅰ. ①宅… Ⅱ. ①深… Ⅲ. ①住宅—建筑设计—作品集—台湾 Ⅳ. ① TU241

中国版本图书馆 CIP 数据核字 (2017) 第 158204 号

出版：中国林业出版社
（100009 北京西城区德内大街刘海胡同 7 号）
http://lycb.forestry.gov.cn/
电话：（010）8314 3518
发行：中国林业出版社
印刷：深圳市雅仕达印务有限公司
版次：2017 年 7 月第 1 版
印次：2017 年 7 月第 1 次
开本：235mm×335mm，1/16
印张：20
字数：300 千字
定价：428.00 元 (USD 86.00)

PREFACE
序言

Design Creates High Quality Life
设计，创造优质生活

Design is like a sketch, each drawing of which can make the paper of the space more exquisite. The quality of life is the foundation to design the space. Designers should enrich themselves constantly and experience life attentively, uphold people-oriented home concept, integrate warmth and comfort into the space and constantly create new aesthetic home design. Design concept and innovation combine together and collide a gorgeous sparkle of new elements. When planning the residential design, we must consider elements such as house orientation, structure, positions of doors and windows, conduct the design according to layout of the circulation and users' required space scale, and understand the owner's living requirements and habits in every communication with the owner so as to design the overall space planning. The design that I want is not design for design but design which combines design purpose with application.

In fact, whether it is self-built house, new house or old house renovation, every different pattern and space design can inject new vitality into space. For self-built house, we cooperate with the architect; in addition to reviewing laws and regulations together, we also consider basic engineering structure and repeatedly discuss and think about the position of pipeline reservations of water and electricity. We also take the utilization which the user didn't concern into consideration, so the self-built house is the best type to conduct the design, because we can customize the expected pattern and requirements and collocate with courtyard landscape design planning to supplement each other so as to become a unique residential design. As for new house and old house renovation, because of the original space pattern, aspects that need to be reviewed would be all-round. Sometimes the design indeed affects the whole plan, so you need to break the original space pattern to meet the owner's requirements and the planning of the storage function.

We hope that through solid practical experience and engineering quality control, we can achieve the expected space works in every owner's inner heart and can witness the owner to enjoy the high quality life, which is the fountain of all my enthusiasm and motivation to design!

设计就有如描绘素描般，每刻画一笔，便能使空间的画纸更添精致。生活质量方为设计空间之本，设计者更应每时每刻充实自我、用心体会生活，秉持着人本为家的概念，将温馨与舒适融入空间，不断创造出新美学的家设计。设计理念与创新结合撞击出新元素的绚丽火花，在规划住宅设计时，我们须考虑房子的座向、结构、门窗位置等因素，在动线格局上的规划及使用者所需要的空间大小进行设计。在跟屋主每一次的会谈中，了解屋主生活使用上的需求及习惯，进而去做整体的规划。我所希望的设计，不是因为要设计而设计，是要和设计目的、用途去做结合。

其实不管是自地自建、新屋亦或是老屋翻新，透过每一次不一样的格局及空间设计，都能让空间注入新的生命力。以自地自建来说我们会与建筑师合作，除了一同检讨法规类之外，基础工程的结构也是不容马虎，水电方面的管线预留的位置也要再三讨论及思考。我们还要为屋主考虑屋主所没顾虑到的使用层面，因此自地自建是最好发挥的类型，因为可以量身订做想要的格局及需求，搭配庭院景观设计的规划，相辅相成而成为独一无二的住宅设计。而新屋及老屋翻新因为有制定的空间格局，所以需要检讨的层面也比较全方面。有时设计牵一发而动全身，所以需要打破制定的空间型态，以达到屋主的需求及收纳机能上的规划。

我们期盼通过扎实的实务经验与工程质量的控管，实现每一位业主心中日夜冀盼的空间作品，能亲眼看见业主抱持着享受去体验优质的生活质量，即是我对于设计保持热忱及动力的一切之泉源！

WYSON DESIGN / Wyson
怀生国际设计 / 翁嘉鸿

CONTENTS
目录

NATURAL STYLE
自然风

LIGHT LUXURIOUS STYLE
微奢风

自然风
NATURAL STYLE

Design Concept
设计理念
Decorative Materials
装饰材料
Natural Colors
自然色彩
Natural Lighting
自然采光
Interior Virescence
室内绿化
Bringing Scenery Into House
引景入室
Space Planning
空间规划
Lighting Design
照明设计

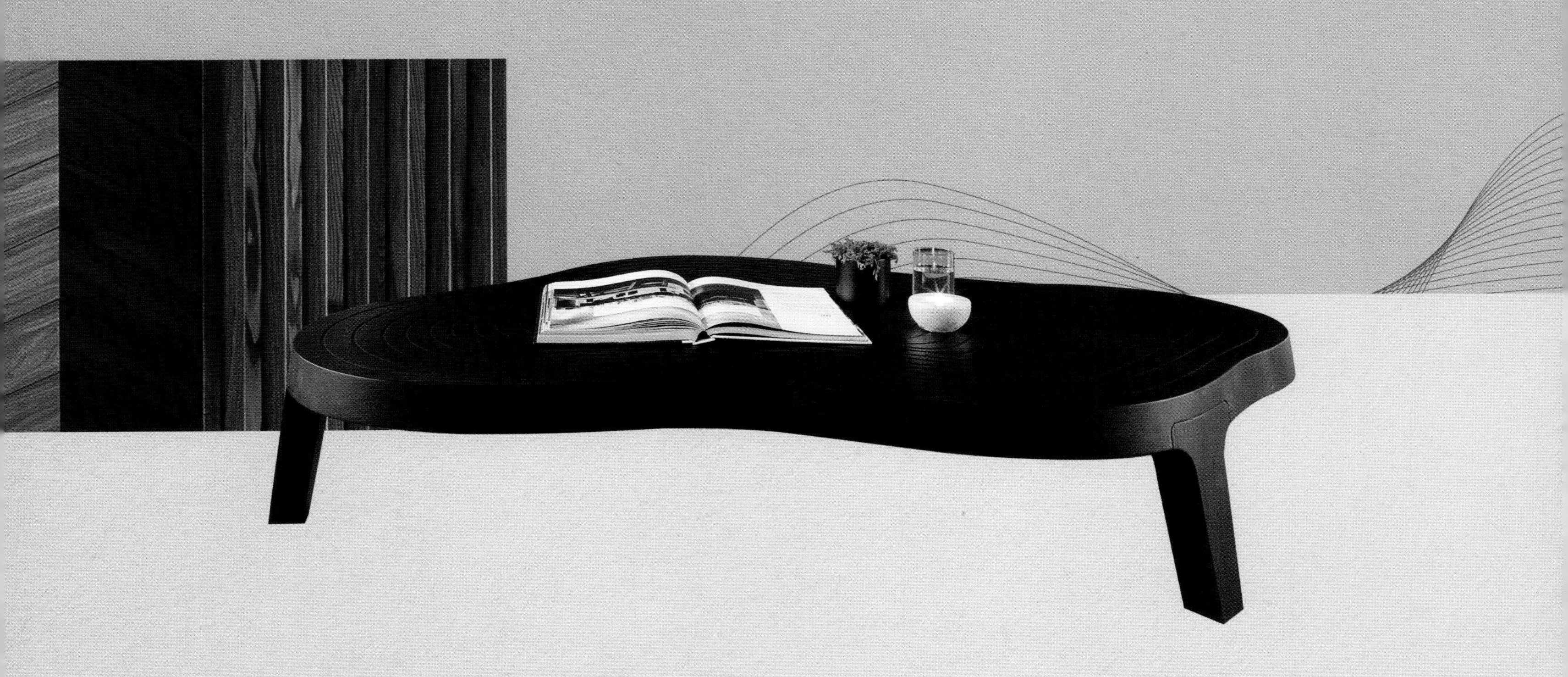

隐藏与探知
Concealment and Exploration

设计公司：天坊室内计划
Design company: TINE FUN Interior Planning Ltd

设 计 师：张清平
Designer: Qingping Zhang

项目地点：台湾台中
Location: Taichung, Taiwan

项目面积：138m²
Area:138m²

DESIGN CONCEPT
设计理念

Designs are developed around the central study. Composed colors and white plain form the purest existence in the space. The partition axis of the space is neat and clear. Crisscrossed structure lines and classic elements bring people the feeling of wanting more. Virtual and real things bring out the best in each other. The designer breaks the inherent mode of the space structure to form a new segmentation and to meet different needs of different people.

The design is guided by literary breath and is endowed with an inherent space temperament which is a heavy aura. Lead residents enter into the established world and let themselves to discover. "Concealment" is used to "explore" better life.

环绕中心书房展开设计，沉稳的色彩和白色的质朴构成了空间里最纯粹的所在，空间的分割轴线利落明朗。纵横交错的结构线条与经典元素带给人意犹未尽的感觉。虚虚实实、相得益彰。打破固有的空间结构模式，形成新的分割，同时也满足了不同人群的不同需求。

以书卷气息为设计主导，赐予一种与生俱来的空间气质，那是一种浓烈氛围的气场。引导居住者进入到既定的世界里，自己去发现；通过“隐藏”来引导“探知”生活的美好。

◇ SPACE PLANNING AND NATURAL LIGHTING 空间规划、自然采光

Large area of glass window promotes the indoor and outdoor air convection, makes sunshine and air circulate naturally, breaks the psychological limitation and creates a transparent freedom. It seems to virtually reflect the value of the design which can highlight the practicability and constitutes a new lifestyle under the premise of using appropriate elements.

大面积的玻璃窗，促进室内外空气的对流，让阳光与空气自然流通，同时打破了心理的局限，营造出一种通透的自在。这似乎也于无形中折射了设计的价值所在：既可以突出实用性，又能在恰当运用元素的前提下，构成一种新的生活方式。

◇ LIGHTING DESIGN　照明设计

Flowing lines endow the space with more comfortable aesthetics. The hidden lamp tubes and light strips fit the whole design. And the manifold droplight in the dining room forms a good interaction with the living room structure, echoing and connecting with each other.

流动的线条赋予空间更多的自在美感，隐藏的灯筒与灯带贴合整体设计。而流带形的吊灯，与客厅结构形成良好的互动，互相呼应又彼此连结。

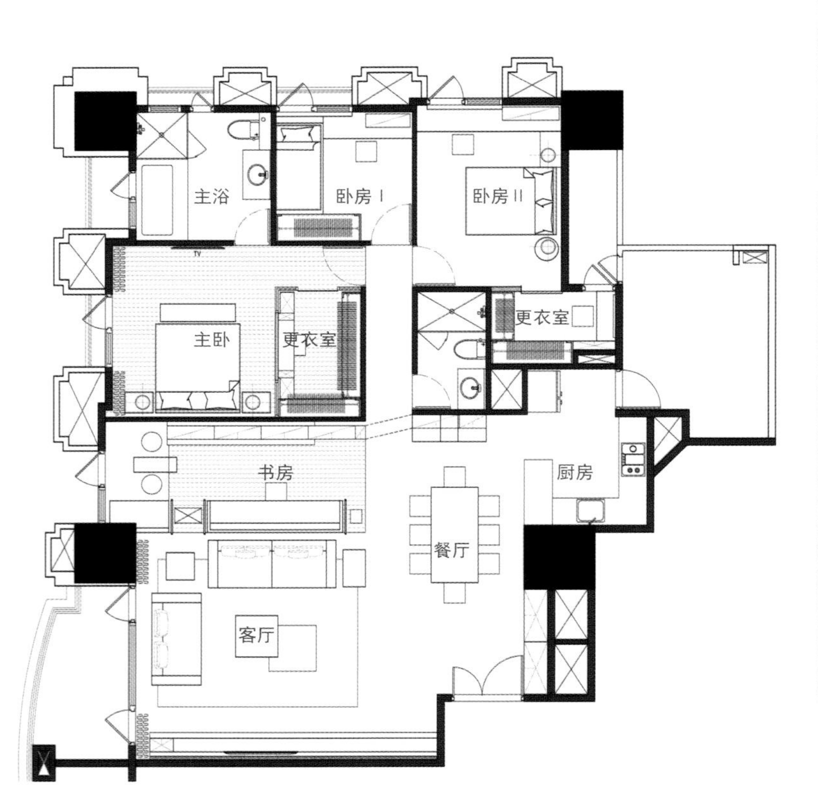

Arletta
Arletta

◇ NATURAL COLORS 自然色彩

Jacinth is dotted in the space and forms a fresh contrast, giving people a surprising sense, which is fashionable, deduces modern conciseness into extreme and echoes with the literary breath. From the point of art, it can also create another aesthetics.

利用橘红色点缀空间，形成鲜明对比，给人豁然开朗的感觉，亦不失时尚，将现代的简洁发挥到了极致、同时也呼应了书卷气息。从艺术角度而言，也能谱出别样的审美层。

CENTURY

露宿自然原野的生活
Living a Natural and Wild Life

项目名称：观山云起
Project name: Closely Viewing Mountains and Clouds

设计公司：怀生国际设计有限公司
Design company: WYSON DESIGN

设 计 师：翁嘉鸿
Designer: Wyson

项目地点：台湾新竹
Location: Hsinchu, Taiwan

项目面积：600m^2
Area: 600m^2

主要材料：铁件、木皮、水泥粉光、石材等
Main materials: iron, wood veneer, cement powder light, stone, etc.

摄 影 师：朴叙空间创意有限公司
Photographer: MD PURSUIT

DESIGN CONCEPT
设计理念

The owner who loves nature farms and reads in the highland, enjoys the bright sky with cotton clouds in the daytime and galaxy stars and moon at night, drinks a cup of tea or gathers and barbecues with families and friends in the breeze, which is infinite happiness. With independent life field and spacious space, the designer uses modern architectural vocabularies to pile stereo feeling of the space so as to create a leisure and original enjoyable circumstances. In the green grass surrounded by trees, the long path links two white plies in different sizes. The small one is planed to be a modern gatehouse as well as a glass reception hall outside. The surface of the solid big one is sewn with lattice to divide proportion of the side elevation and to lighten the burden, while the front elevation uses transparent clear glass and slight iron to outline detailed structure, to present the real and virtual contrast of deign technique and to capture the good weather of blue sky and white clouds.

热爱大自然的业主，耕读高地，享受白天的晴空万里与棉花云朵、夜间的银河星斗与高挂明月，微风吹抚中，品茗一杯或是与家人朋友团聚烤肉，欢乐无限。遗世独立的生活场域，空间阔绰，设计师运用现代建筑语汇，层层堆砌空间中的立体感，营造休闲且匠心独运的写意情境。大树围绕的青葱草地上，悠悠路径，链接两处堆栈成型的大小白色量体，小量体规划为现代感警卫室，同时也是对外的玻璃屋交谊招待厅。札实的大量体表皮上，施工缝以格子状分割侧立面比例，减轻负担感，如轻触大地，正立面则装上晶透的清玻璃与纤细铁件勾勒细节轮廓组构，展现对比设计手法上的实虚，也反射捕捉了户外蓝天白云的好天气。

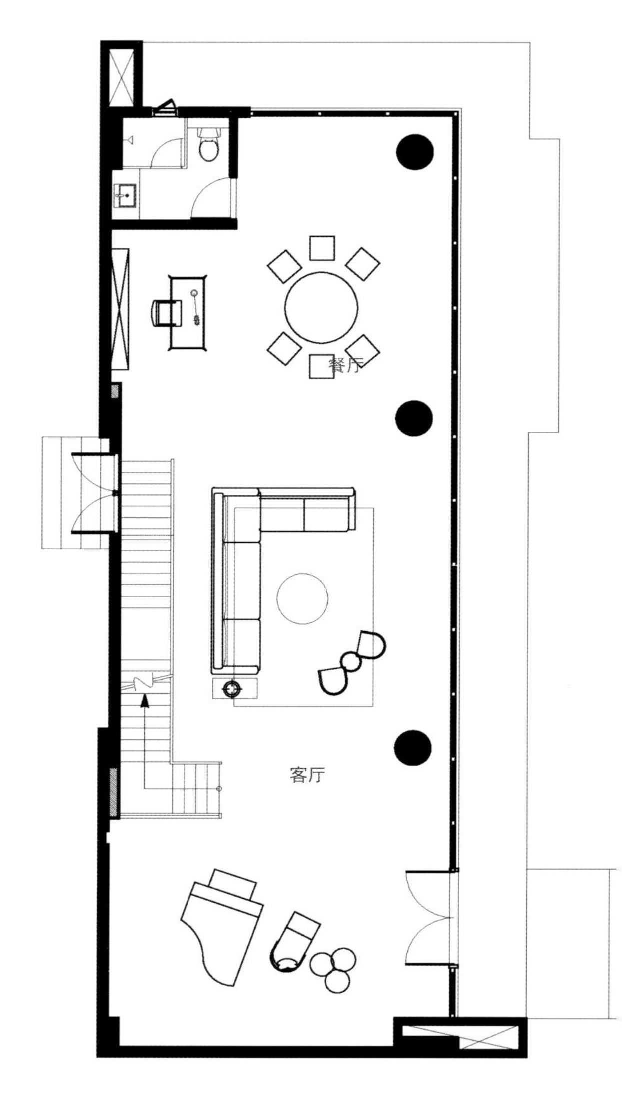

◇ SPACE PLANNING 空间规划

The internal path and public and private areas are divided by the tall and special dark L-shaped ladder in the center. If you want to enter into the absolute private domain, you need to walk through the public area in the first floor and go to the second floor step by step through the stairs near the solid wall. Walls of the master bedroom are not over decorated and keep the original gray scale of the architectural concrete structure. Because of the regional independence of the building, lager pieces of window have no influence on the owner's privacy and bring in a lot of natural daylight so that the entire second floor is bright and comfortable. The log furniture with vigorous modeling and warm materials, abstract paintings on the concrete wall, velvet carpet and gold bedside lamps together soften the hardness of the space.

内部路径与公私领域，由平面中央高耸而特殊的深色 L 型阶梯区分高程破题点出，欲想进入绝对的私密领域，得透过步伐穿过一楼的公领域，通过一层层攀登崁进实墙边界的踏阶至二楼。主卧里的天地壁不过度装修，留下建筑混凝构造的原始灰阶，也因建筑物的区域独立性，大面开窗并不影响业主的隐私，反而引入了大量的自然天光，整体二层空间显得明亮宽适。造型霸气而材质温润的原木家具，与混凝墙面挂上的抽象画作与绒毛地毯、金色床头灯，一起柔化空间的硬度。

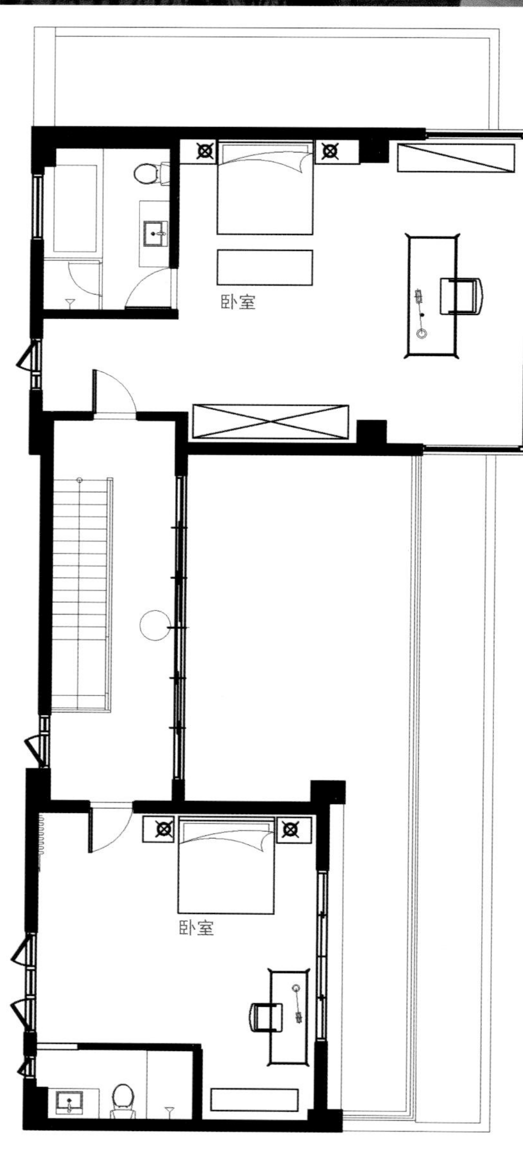

◇ DECORATIVE MATERIALS 装饰材料

The five-meter first floor makes the private area spacious and decent. The settling ceiling with frame block customized lamps and dark main wall flagstone presents courtyard images and upward three-dimensional sense. In the spacious space, thick and large structure podetium contrasts with the lightsome stairs, presenting the vertical dimension and infinite imagination with distinct tension.

底层五米的挑高手法，让居家的公领域宽敞大气，安插沉降天花板，围塑框架块体订制灯与深色主墙石板，转绎了天井意象与向上突破的立体层次感，在诺大空间中厚实大结构柱体对比于轻盈的阶梯，皆展示贯穿垂直向度，别具张力的无限想象。

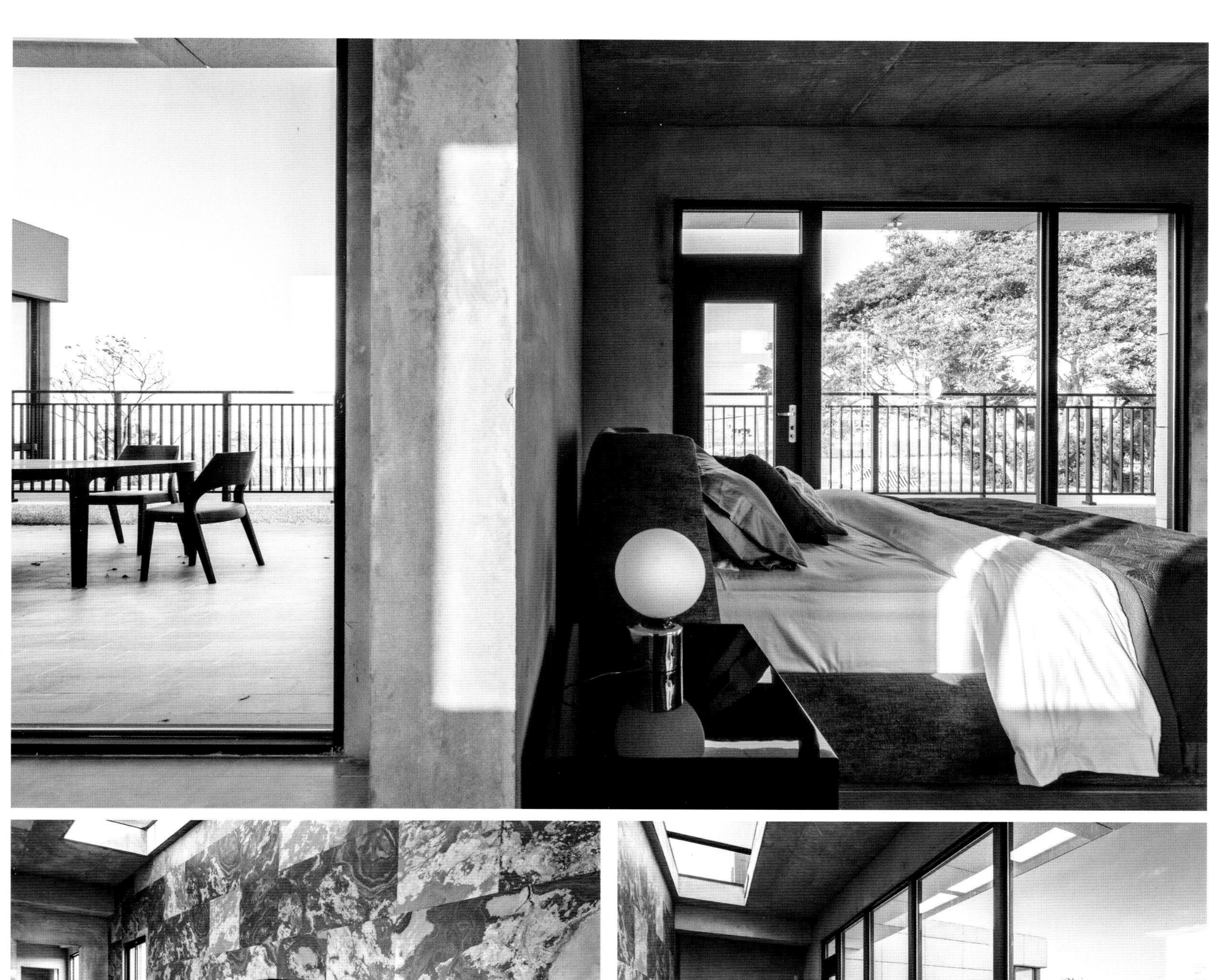

◇ BRINGING SCENERY INTO HOUSE 引景入室

The interior view uses the layout strategy of two sides open to the courtyard. The symmetric small windows on the left closed wall are placed with abstract art paintings. It equally divides the planar proportion of rectangles in different sizes, which properly adjusts the privacy of the space and adds temperament. The ink-colored stair and relief main wall bring the theme of the space. The open front elevation and right side elevation made of glass frame the outdoor garden scenery. Under the physic effect of light and glass, the tree truck and cylindrical mirror create an interesting effect in front of the dining table. The water line and the landscape pool are built around the building, integrate the water field and half of outdoor space and handle the intermediary of the indoor and outdoor spaces. Step into it or throw a stone into it, it can wash the hustle and bustle.

屋内视野，采取两面对庭院开放的布局策略，左侧封闭实墙上的对称小窗口中间放置抽象艺术画作，均分了大小矩形的面状比例，在适度调整空间隐私性的同时增添气质，而浓淡墨色的楼梯与浮雕主墙则带出空间主题。大面玻璃开放的正立面与右侧立面，框景户外园艺，在光线与玻璃的物理作用下，餐桌前，树干与圆柱的间隔镜面延伸效果，相当有趣。落水线和景观水池围绕建筑，整合水域与半户外空间，处理了空间内外的中介，踩踏入口跳石入内，洗涤尘嚣。

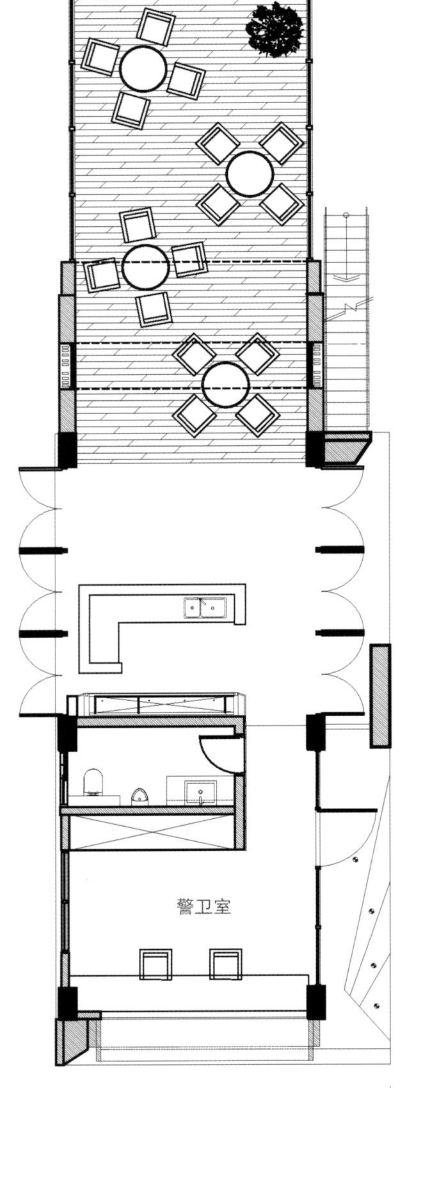
警卫室

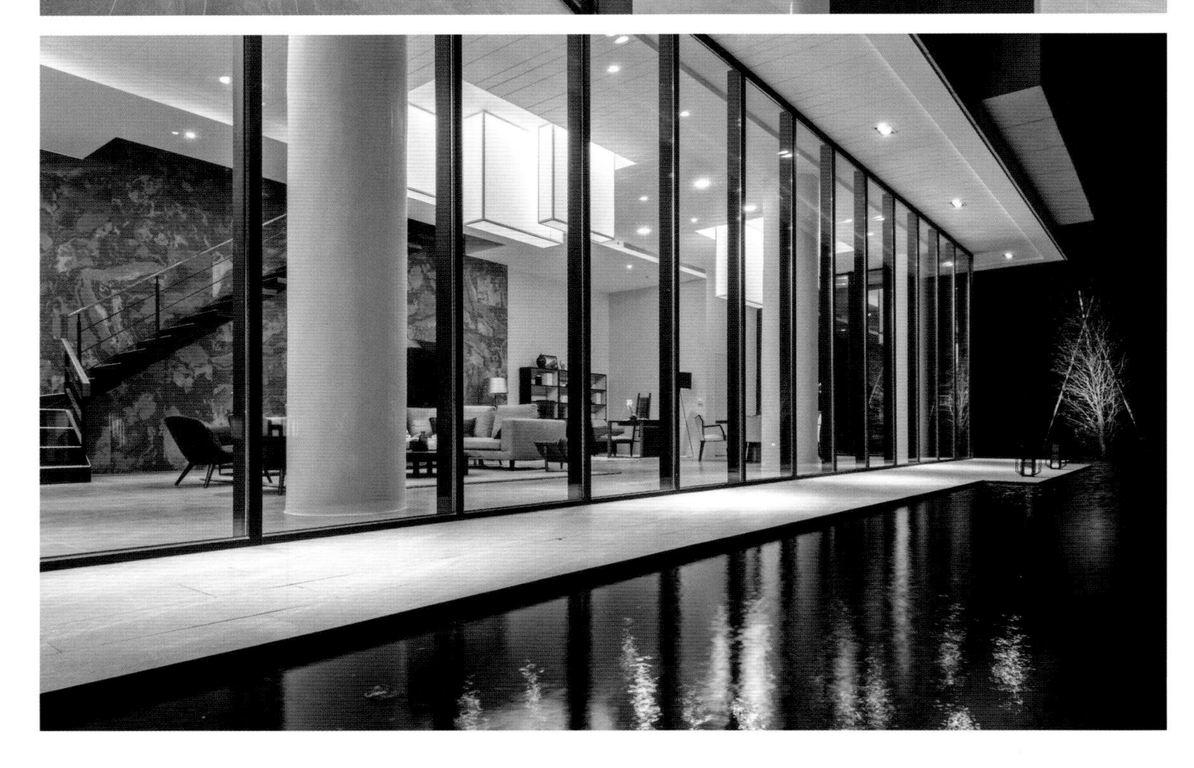

光影共舞 质朴之邸

Light and Shadow Dance Together in the Plain Residence

设计公司：全育空间设计有限公司
Design company: TONGYU Space Design CO., Ltd.

设 计 师：庄媛婷、郑瑞文
Designers: Chuang Yuan-Ting, Cheng Jui-Wen

项目地点：台湾台南
Location: Tainan, Taiwan

项目面积：740m²
Area: 740m²

主要材料：钢刷白橡木皮、天然胡桃木皮、进口壁纸、茶镜、裱布等
Main materials: steel brush white oak wood veneer, natural walnut wood veneer, imported wallpaper, tawny glasses, cloth, etc.

摄 影 师：许启裕
Photographer: NIKE

DESIGN CONCEPT
设计理念

The house uses as-cast-finish mould to present its appearance, complemented by warm wood to alleviate the cold tone produced by cements. For example, in the open public field, the designers intentionally use some solid wood objects in a large number of the original stone wall, integrate hotel Lobby design concept and inject fashion vocabularies of art gallery to create an elegant and composed space.

Appropriate space atmosphere which can stabilize body and mind, thinking of beautiful natural scenery, are the effects this project aims to create. The designers use earth tone to heal the soul and wood texture to highlight space layering. Large piece of French window brings in outdoor waterfall and grille to make light and shadow skillfully outline the atmosphere expression. The charming green window scenery adds a tranquil natural breath for the space.

家屋以清水模形貌呈现，为了缓和水泥散发的冷凝调性，则辅以温润木作面材，例如开放公场域刻意使用些许实木物件在大量原石墙中缀点，并融合饭店 Lobby 设计概念，穿戴艺廊的时尚语汇，内敛成馨雅沉稳的空间。

适当的空间营造能稳定心灵，想着大自然让人流连忘返的意景，这便是此案想创作出的效果。运用大地色系作为疗愈心灵的基调，再以木作肌理凸显空间层次，而大片落地窗延揽了户外水幕墙及格栅，让光影巧妙勾勒氛围表情，导入动人绿意的窗景，便为空间凝塑出静谧的自然气息。

◇ SPACE PLANNING 空间规划

The five-story residence has delicate and complete space planning. The public and private areas are separated with clear priorities, making life of the family go on orderly every day. Entering from the porch in the first floor, you can see the Zen-like reception hall which can accommodate some friends to chat and taste tea here. The public area in the second floor is divided by the central stair with the living room in the left and dining room in the right. The open kitchen adds interacting fun for cooking. The entire third floor is the private area of the owner with master bedroom, dressing room, bathroom and SPA, to maximally give the owner the best quality life. In the fourth floor, the front and back bedrooms with independent bathrooms are symmetrical distributed by the stair. In the fifth floor, there are guest room, wine tasting room and gym, providing places for residents to relax and taste life.

五层住宅规划细致完善，公私分开，主次分明，让一家人的生活每天都有秩序地进行着。从一楼玄关进入室内，富有禅意的接待大厅可容得下三五好友谈笑风生，品茶叙事。二楼公共领域以楼梯为中界线，客厅和餐厅左右分布，开放式的厨房增加了烹饪时互动的乐趣。三楼整层是主人的私人领域，主卧、更衣室、浴室和SPA室等一应俱全，最大化地给予主人最优质的生活。四楼前卧和后卧以楼梯间为界对称分布，并带有独立的浴室。五层除去一间客房的私领域，同时还有品酒区和健身房，给居住者带来休闲和品位生活的余地。

◇ NATURAL COLORS 自然色彩

Colors on interior space can easily affect the mood of residents. Warm color can make people pleasant while cold color can calm people down and wood color is warm and comfortable.The basic tone of this residence is earth color; with elegant gray background wall, wood floor, creamy-white cloth sofa and Volakas, the addition of different colors makes the whole space send out a unique charm.

室内空间的色彩选用往往也容易影响居住者的心情，暖色系使人心情愉悦，冷色系令人安静理智，木色则温润舒适。这个住宅空间则以大地色系作为空间的基础色调，雅致灰的背景墙、木质地板、米白色的布艺沙发、爵士白的大理石等，不同色彩的加入，让整个空间散发出独特的居住魅力。

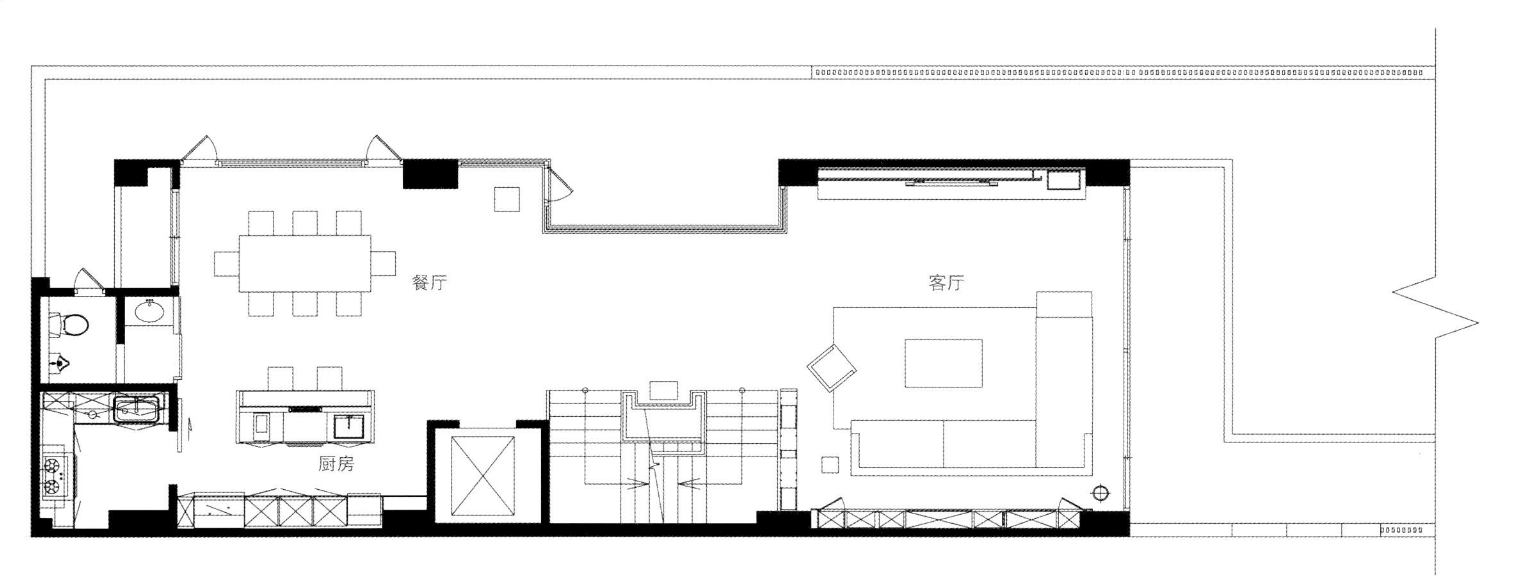

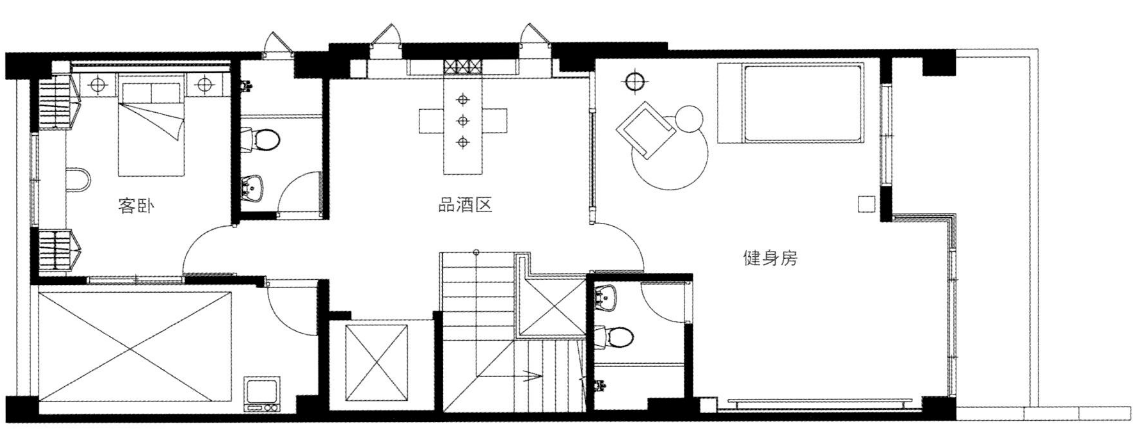
客卧
品酒区
健身房

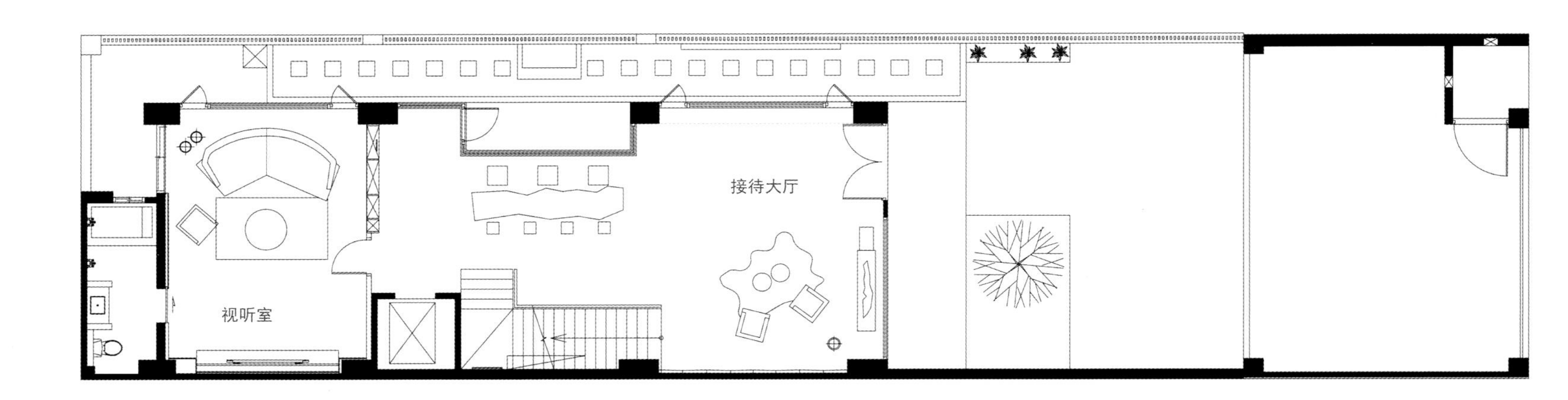
接待大厅
视听室

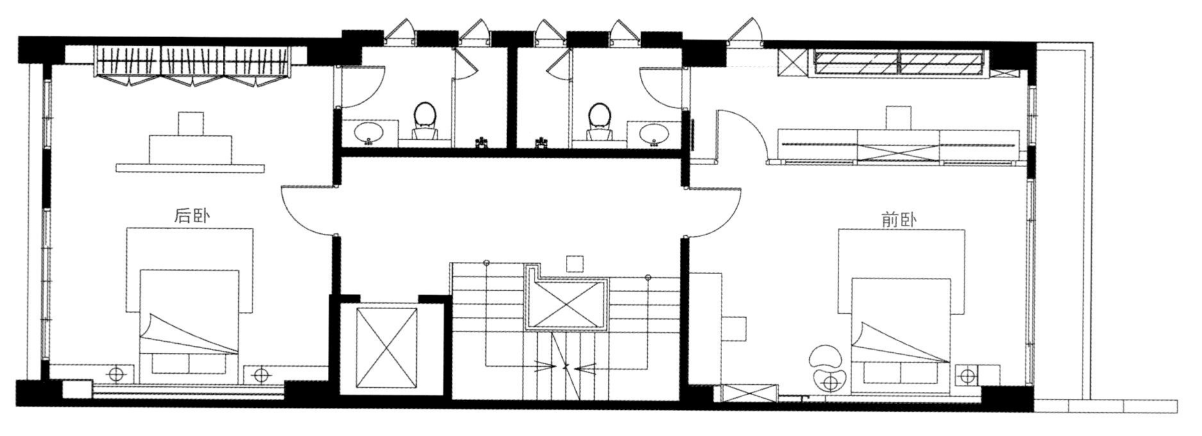
后卧
前卧

◇ BRINGING SCENERY INTO HOUSE 引景入室

The five floors all keep large pieces of French windows, freely bringing light and shadow interior, forming a virtual and actual scenery in the space. Through large area of French window, green scenery outside the window jumps into the eyes naturally, adding a greenery for the space. At the same time, transparent and clean windows also greatly increase the communication between indoor and outdoor spaces, making life extend to outdoor and green return to life.

五层空间都留有大片落地窗，光影自由导入，在空间中构成虚实交错的景致。借由大面积的落地窗，窗外绿意盎然的景色也自然映入眼帘，为室内增添一抹绿意，同时透明洁净的窗户也大大增加了室内与户外之间的空间交流，让生活延续至户外，绿意回归到生活。

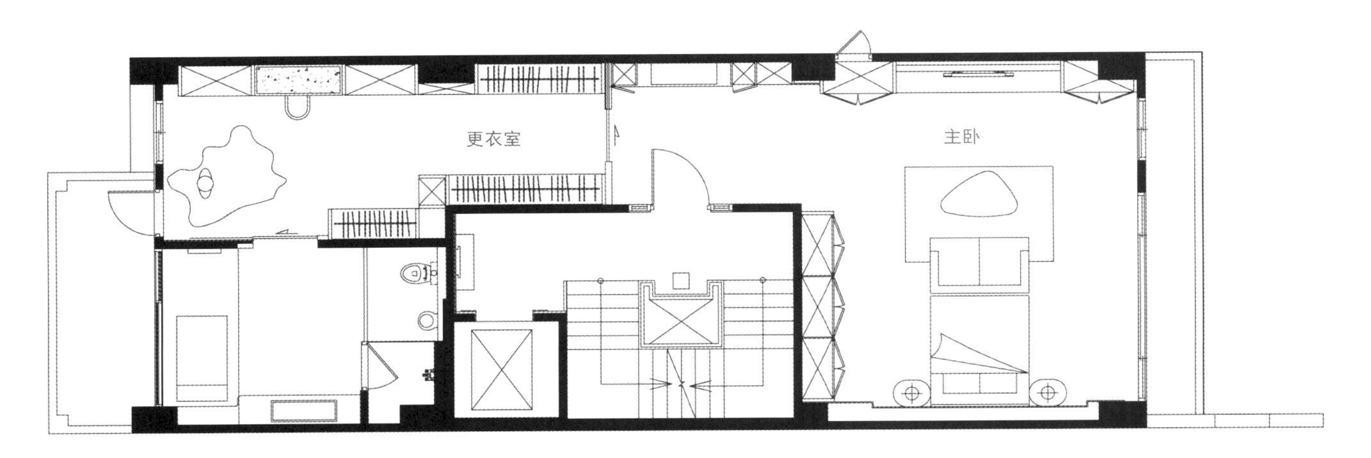

天地合一的海边秘境

Seaside Fairyland with Oneness of Heaven and Earth

项目名称：阿都兰
Project name: A'tolan House

设计公司：创研空间
Design company: Create+Think Design Studio

设 计 师：何俊宏
Designer: Arthur Ho

项目地点：台湾东部
Location: Eastern Taiwan

项目面积：992m²
Area: 992m²

主要材料：现地采集的石头、钢构、混泥土、杉木、玻璃、石材等
Main materials: stone, steel, concrete, wood, glass, marble, etc.

摄 影 师：李国民
Photographer: Kuomin Lee

DESIGN CONCEPT
设计理念

The base is located on the east coast of Taiwan near the Pacific Ocean and is about 90 meters long from south to north in long and narrow shape; the coastline is vertical from north to south to 60 meters and turns to the east in 130 degrees. The east-west width is 30 meters; the contour line falls steeply from west to east with three layers. The designer uses the original method, the A'tolan, builds the walls by stones dug out, from west to east, and piles up three layers from high to low as terraces; the high layer is 28 meters with parking entrance and front yard; the middle is 23 meters with buildings; the lower is 20 meters with tree house and green land.

The charming of this project lies in the clarity of consideration. From living area to the selection of building materials, the architectural designs harmonize and integrate with the structure and surrounding landscape. The interior functional design is also deeply considered and planned. The space planning allows lights and air flow freely in all directions, creating a gentle and comfortable atmosphere.

基地位于台湾东海岸，邻太平洋，呈南北向狭长形，长约90米，海岸线自北垂直向南至60米处向东转角130度；东西宽30米，等高线自西向东陡降，呈三层高度。设计师采用原始工法——阿都兰的工法，以基地内挖出的石头，砌石为墙，自西向东，由高至低如梯田般堆砌出三个层次：高层有28米，为入口车道和前院；中层有23米，主要有建筑物；低层20米，分布有树屋及绿地。

本案的魅力在于包含全方面清晰的考量。从生活面积到建材选择，建筑设计融洽整合结构体以及周围景观。室内功能设计亦经过深思熟虑及规划。空间的规划允许光线和空气四面八方自由地流动，创造了一个和缓舒适的气氛。

◇ DECORATIVE MATERIALS 装饰材料

The building appears in front of the sea in a humble posture, flat and extended, paralleling to the shoreline and embracing the sea along with the shoreline in 130 degrees. The building piled by steel structure is surrounded by stone walls; its roof is high footpath planted with all kinds of herb plants, blurring out the structure of the building by landscape. Downstairs from the big stone stair to pass through the high footpath, the below half-open area is the porch to the main building.

建筑物在大海前以谦卑的姿态出现，扁平而延伸，平行于海岸线，随着海岸线 130 度展开拥抱海洋。以钢构筑而成的建筑物被堆砌的石墙环绕其中，屋顶是种植各式香草植物的高台步道，以地景方式虚化建筑量体，从大石阶拾级而下穿过高台步道，下方的半露天区是往主建物的玄关。

◇ SPACE PLANNING 空间规划

In the need of the shelter of nature, the designer and the team choose the lowest possible way just to satisfy the need of shelter and cooking. Therefore, the interior space of this residence is composed of cooking space and sleeping space that are necessary for life. The residence is structured by glass French doors and windows which can be open and close, where lights and air can flow. Being here, you can face the spaciousness of the sea and can feel the power of the embrace of the big stone walls. Other spaces unnecessary for life use the way to symbiose with nature. The space outside the sleeping area is the open-air bath extended from the skylight shower area. The long table on the side of the half-open porch is the dining area. The outdoor living space in front of the building can be a yoga platform, a star theater or any possible space.

在大自然环绕下的居所需求，设计师和创研团队选择最低度的方式，只需满足遮风避雨藏身和取火煮食即可。于是，此居所的室内空间仅由生活必需之烹煮空间与睡眠空间组成。居所是以可以开合的落地玻璃门窗所组构，光线空气贯穿流动其间，可面向大海的开阔，或感受大石墙厚实环抱的力量。其他非生活必要空间采用与自然共生方式，睡眠空间外是天光淋浴区延伸的露天澡盆；半露天玄关侧的长桌是用餐区；建筑物前方有户外起居空间，可以是瑜珈平台，可以是星空剧院，亦可以是任何的可能性空间。

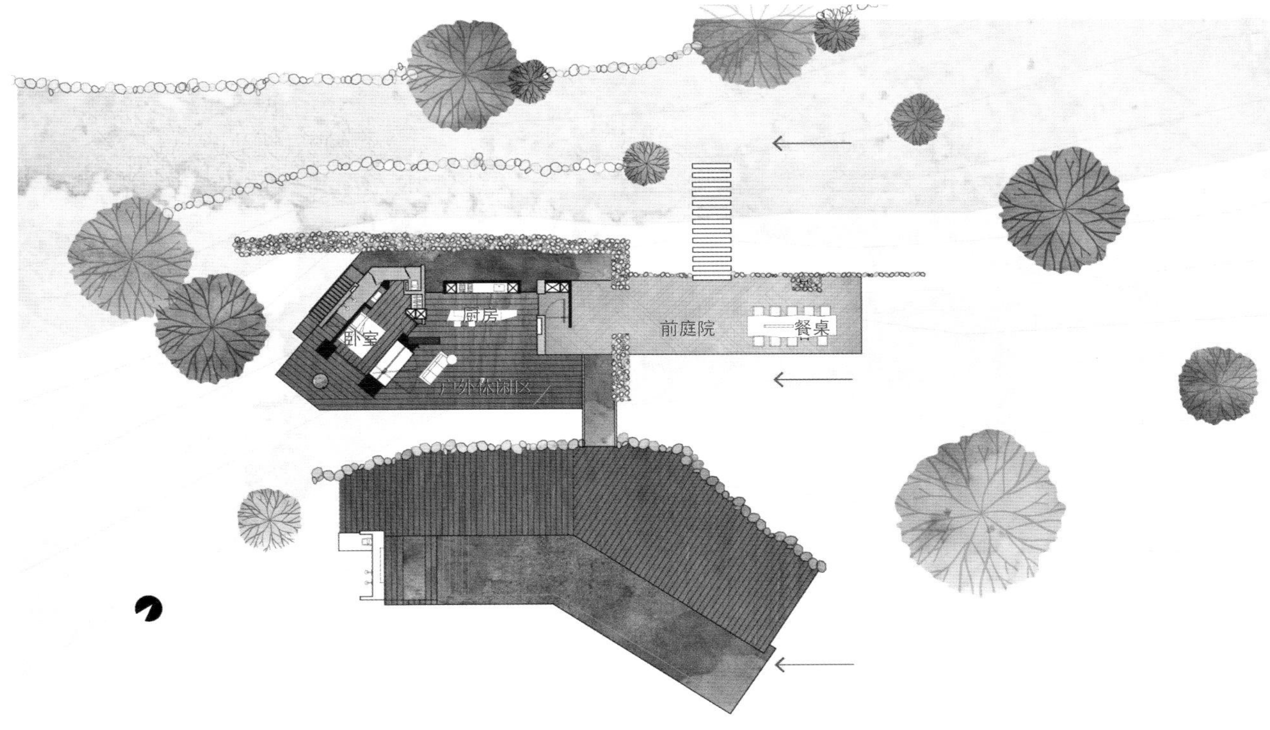
卧室
厨房
前庭院
餐桌
户外休闲区

御·光 Defense and Light

项目名称：麒麟御
Project name: Kylin

设计公司：近境制作
Design company: Design Apartment

设 计 师：唐忠汉、许郁晨
Designers: TT, Yuchen Xu

项目地点：台湾台北
Location: Taipei, Taiwan

项目面积：198m²
Area: 198m²

主要材料：铁件、染色木皮、平光黑网石、绷布、漆面、玻璃等
Main materials: iron, dyed wood veneer, flat black mesh stone, fabric, lacquer, glass, etc.

摄 影 师：岑修贤摄影工作室MW PHOTO INC.
Photographer: MW PHOTO INC.

DESIGN CONCEPT
设计理念

This project uses open pattern design, reduces compartment partition to create visual perception of large scale space, and uses floor material changes to define fields. The living room uses the condition that one single side has lights to design TV wall so as to make light and shadow switch to increase layered changes of wall. The bottom wall in the dining room match the whole space side scene display cabinet with the long dining table and droplight, showing the center of family life and party. Wandering in it, the bottom of the visual focus is a scenery, a favorite object also an extension of memory. Quiet and composed colors are used to elaborate the space structure as a whole, emitting a natural, humanistic and comfortable life atmosphere under the decoration of stone and wood, iron and cloth.

本案采开放格局设计，减少隔间阻隔创造大尺度空间视觉感受，以地坪材质变化界定场域。客厅利用单面迎光条件设计电视墙，使得光影切换增加壁面层次变化；餐厅底墙以整面端景展示柜搭配长形的餐桌及吊灯，展现家庭生活与聚会的中心。游走其间，视觉焦点的底端是一处风景、是一件喜欢的物件，也是一处记忆的延伸。以整体宁静沉稳的色系铺陈空间量体，在石与木、铁与布构筑之下，散发出自然人文的舒适生活氛围。

◇ SPACE PLANNING　空间规划

The living room and dining room separate faintly by the beam of the building. Wood veneer of the beam extends to the wall and connects with the gray lens, bringing out the Western kitchen space. The planning of furniture and island derives out the open rectangular-ambulatory corridor, making the whole public area become a circulatory interactive area. Lots of light and shadow pass through wood shatters and can flow freely to perform time of light.

客厅与餐厅之间借建筑大梁隐约分界，梁上木皮延伸到壁面接续成灰镜，带出西厨空间，运用家具及中岛的规划衍生出开放式回字型信道，让整个公领域变成一个循环的生活互动领域。大面光影透过木百叶，便能恣意流动表现光的时间。

◇ NATURAL LIGHTING AND NATURAL COLORS
自然采光、自然色彩

The master bedroom is given priority to light gray tone. The head of the bed uses simple soft fabrics to match with gray fluorhydric acid glass which corresponds to the room entrance. Desk lamps form a visual side view to link the inside and outside corridors. The room deliberately separates from the window edge to form a corridor, bringing lights in interior to create a rest area.

In the elders' room, the desk is the bottom view which connects the corridor of the public area with the bedroom of private area. Opening the door is an extension of the space while closing the door is a part of the room. The space configuration like a side view softens the absolution of the pattern diffusion and becomes a unique restrained and tranquil feature under the foil of glorious lights.

The children's room extends the public area floor into interior to be included in the a bathroom area. The sliding doors create a bilateral using space. The transition of areas and the desk island extend and define the sleeping area. Wall display cabinet matches warm gray with white to make it better manifest users' personalities.

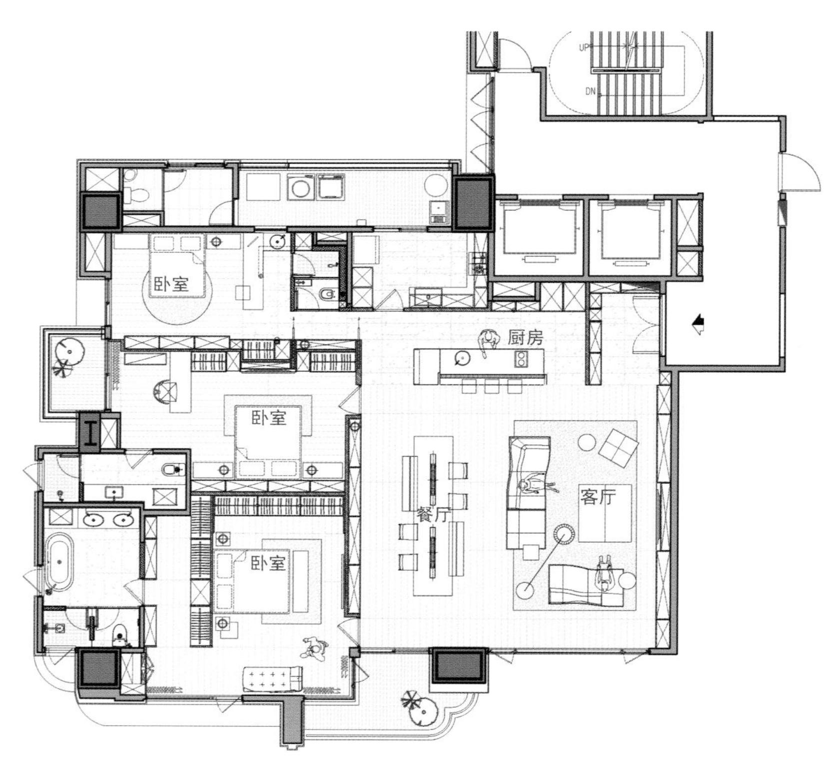

主卧室以浅灰色系为主，床头采用简约软质绷布分割搭配灰色氟酸玻璃，氟酸玻璃对应到房间入口位置，台灯形成链接内外廊道的视觉端景；房间刻意退开窗边形成廊道，导引光线入内创造出卧榻休憩区。

长辈房书桌亦是公领域廊道接私领域卧室的底景，开门是空间的延伸，关门是房间的一部分，端景般的空间配置，柔化了格局划分的绝对，在洒落的光辉照映之下，成为独特的内敛宁静特质。

儿童房将公区地坪延伸入里，交错形成卫浴范围，运用拉门开合创造双向的使用空间。地坪的转换以及书桌岛台延伸界定出睡眠区域，壁面陈列柜体用暖灰和白色色调错置搭配，让展示空间更能彰显使用者个性。

大隐·静寂
Great Seclusion, Tranquility

设计公司：大湖森林室内设计
Design company: Lake Forest Design

设 计 师：柯竹书、杨爱莲
Designers: Zhushu Ke, Ailian Yang

项目地点：台湾台北
Location: Taipei, Taiwan

项目面积：135m²
Area: 135m²

主要材料：石材、实木、玻璃、铁件、水泥粉光地砖、木地板等
Main materials: stone, solid wood, glass, iron, cement powder light tile, wood floor, etc.

摄 影 师：林福民
Photographer: Fumin Lin

DESIGN CONCEPT
设计理念

The host who is engaged in financial industry, is highly interested in integrating high-tech products into their lives and doesn't want it too cold without warmth of home. And the hostess wants to have a lot of hidden storage space. So they find Lake Forest Design for help.

In order to integrate the host's high-tech audio-visual equipment into space and meet the hostess's need of strong storage, the designers skillfully use all kinds of different scales of segmentation and integration to digest and integrate functions in the space. For example, the wood ceiling of the living room combines with irregular sound-absorbing materials to design ceiling modeling with different functions so as to integrate functions such as speaker, lamp trough and sound-absorption to obtain perfect sound quality, image and sound channel of the audio-visual equipment. The main wall of the living room uses stones in light, moderate and dark gray, combines rational and rhythmic line structures with wood wall elevation modeling to shape the tranquil forest image and texture atmosphere at the same time to integrate concealment of storage function and opening. The temperature is put in the gray and cold space to shape tension and expression of it in the disordered rhythm.

从事金融业的男屋主，因为对于高科技产品融入生活有着高度兴趣，但又不想太过冰冷，失去家的温暖度，再加上女主人希望能拥有大量的隐藏式收纳空间，于是找到大湖森林室内设计协助处理。

设计师为使男屋主的高科技视听设备融入空间里，同时又能满足女主人对于强大收纳的需求，因此巧妙地运用各种不同尺度的分割与整合，去消化融合空间里的机能。例如客厅的木质天花结合错落的吸音材质，设计出不同机能的天花造型，也将喇叭、灯槽、吸音等功能整合于其中，让视听设备的音质、影像及声道完全不打折。并在客厅主墙选用浅、中、深三种灰阶石材，以理性律动的线条结构，结合木构墙体立面造型，形塑宁静的森林意象与自然的肌理氛围，同时整合收纳机能与开口的隐没，将温度置入灰阶冷调的空间，更在错乱的律动中，形塑出空间张力与表情。

◇ DECORATIVE MATERIALS 装饰材料

The space uses cement powder light tiles to warm wood floors and defines function attributes of the field at its convenience. At the same time walls are injected with vertical lines and wood veneer natural grains to expand the decoration and vision. Concave exhibition space is arranged in the wall, which outspreads the vision through the mirror background, lengthens the depth of view and uses display art works to present life tastes of residents.

空间中以水泥粉光地砖拼接温润木地板，顺势界定场域的机能属性，同时于墙面加入垂直线条语汇与木皮天然纹理，达到装饰与视觉扩张作用。且在墙面规划内凹展示空间，透过镜面背景铺陈延伸视野，在拉长景深同时，也运用陈列艺品展现居者的生活品味。

◇ NATURAL LIGHTING　自然采光

The designers skillfully set a oblique edge under the mirror in the original bathroom without good lighting and ventilation, which integrates the dry and wet area, meets different functional properties and keeps the opening of the window to make the bathroom become the most beautiful and natural art performance space.

设计师将原本采光通风不佳的浴室空间，透过巧思在浴室镜面下缘设置一道斜切，不但整合了干湿区域，也满足了不同的机能属性，并保留了对外窗的开口，让浴室空间成为最美最自然的艺术展演空间。

◇ SPACE PLANNING 空间规划

As for the space layout, the designers open up the original compartment of the living room, dining room and study to form a spacious public space, to bring outdoor lighting and landscape into interior and to make people feel enough natural lights dancing in the room when walking in it. Furniture is used to define the space field, for example, the half-high special stone made desk defines the living room and study, creating a transparent and coherent vision. What's more, the stone island is embedded into the dining table between the dining room and kitchen to define interior and exterior functions and activity area. Under the reflection of mirror, stone and lamplight, it reflects different space expressions like glistening light of waves and shadows.

在空间格局方面，将原本客厅、餐厅及书房的隔间打开，串连宽敞的公共空间，也使户外的采光及景观得以入内，让人行走其中都可以感受到充足的自然光源在空间里舞动着。并运用家具界定空间场域，如在客厅、书房之间半高的特殊石材量体制作书桌作为界定，创造通透连贯的视觉。以及在餐厅与厨房之间，运用石材制的中岛嵌入餐桌，作为界定内与外的机能及活动区域等。在镜面及石材、灯光的反射之下，如波光倒影般呈现出不同的空间表情。

暮光合影 共鸣之处

Twilight and Shadow in Resonant Place

设计公司：全育空间设计有限公司
Design company: TONGYU Space Design CO., Ltd.

设 计 师：庄媛婷、郑瑞文
Designers: Chuang Yuan-Ting, Cheng Jui-Wen

项目地点：台湾台南
Location: Tainan, Taiwan

项目面积：586m²
Area: 586m²

主要材料：天然胡桃木皮、天然红橡木皮、普罗旺斯石材、进口壁纸、铁件、灰玻、黑镜、烤漆等
Main materials: natural walnut wood veneer, natural red oak wood veneer, Provence stone, imported wallpaper, iron, gray glass, black mirror, stoving varnish, etc.

摄 影 师：许启裕
Photographer: NIKE

DESIGN CONCEPT
设计理念

This project is based on modern fashionable style, emphasizes on modern, simple, unique and leisure tones and injects artistic conception of natural concept into space design. The natural landscape concept creates new space functions to emphasize on liquidity and openness of the space and to pursue the variability of multiple reorganization and elastic space of functions. It has rich layering and is natural everywhere. At the same time, it uses concise and fashionable design languages to integrate complex space functions so as to pay attention to comparative richness and beauty.

Living here, the owner can enjoy the comfortable and concise texture brought by modern decorations and also can enjoy the leisure of natural breath. Numerous natural landscapes, such as courtyard, flowers and plants, roof garden and balcony in every floor, strengthen the exchange between indoor and outdoor spaces and also make greens of man and nature integrate and symbiose harmoniously.

此设计案是以现代时尚风为设计主轴，强调现代、简约、独特与休闲感等调性，亦将自然概念之意境注入空间设计中。运用自然中的景概念创造新空间机能，以强调空间的流动性与开阔性，并讲求机能的多元重组与弹性空间的变化性。层次丰富，无处不自然，同时运用简约时尚的设计语汇，整合复杂空间机能，借此注重对比的丰富性与美感。

业主居住在此，既可以享受现代装饰带来的舒适简约质感，也能享受自然呼吸的惬意。众多自然景观的加入，庭院花草，天台花园，每层都有的阳台加强室内外之间的交流，也更加让人与自然的绿意相互融合，融洽共生。

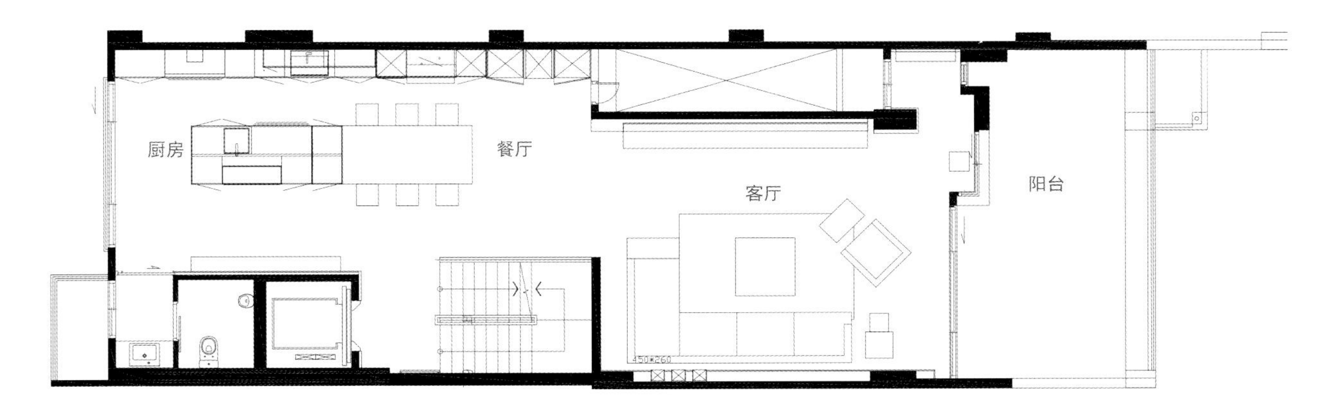
厨房
餐厅
客厅
阳台

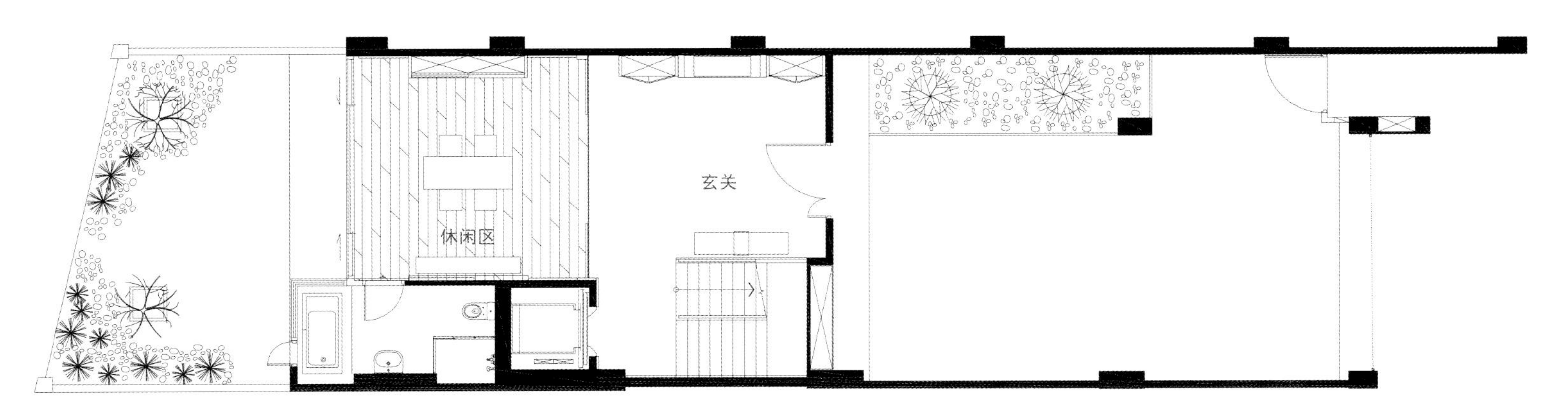
玄关
休闲区

New colors

◇ SPACE PLANNING 空间规划

This project has five layers and every floor is independent and connects by the stairs; the public and private spaces are separated. Whether you sit alone quietly or chat widely with the guests, all can give you enough space to slowly enjoy the comfort and beauty brought by life. There are garage, entrance porch and Zen-like leisure room in the first floor. The courtyard landscape is as unique as the artistic conception that "the flowers and plants in the abode are exuberant". In the second floor, the living room and kitchen are connected, strengthening communications between fields. There are master bedroom, dressing room, bathroom and balcony in the third floor, providing the most convenient and comfortable life for the owner. The exquisite jewelry room displays the hostess's favorite things. There are two bedrooms in the fourth floor, the female bedroom and the front bedroom with independent cloakroom and bathroom. The fifth floor is tranquil and peaceful with a guest room and a roof garden; the nearby audio-room also adds fun occasionally.

本案有着五层住宅空间，每层之间相互独立，公私领域相间隔开，楼梯之间搭建起各层的联系。不管是一人独自幽坐，还是会客聊天间的天南地北，都能给你足够的空间细享生活带来的每一份惬意和美好。一层有车库、入门玄关和禅意休闲室，庭院造景别有洞天，“禅房花木深”的意境大概便是如此。二层的客厅和厨房相互连通，加强场域之间的对话交流。三层是主卧套房，更衣室、浴室、阳台等，为屋主提供了最为便利舒适的生活，精致的珠宝室陈列着女主人的心爱之物。四楼有两间卧房，女卧和前卧同样也有独立的衣帽间和浴室。五层环境安静清幽，有一间客房，屋顶花园在此相伴，一旁的视听室也增添了偶尔间的乐趣。

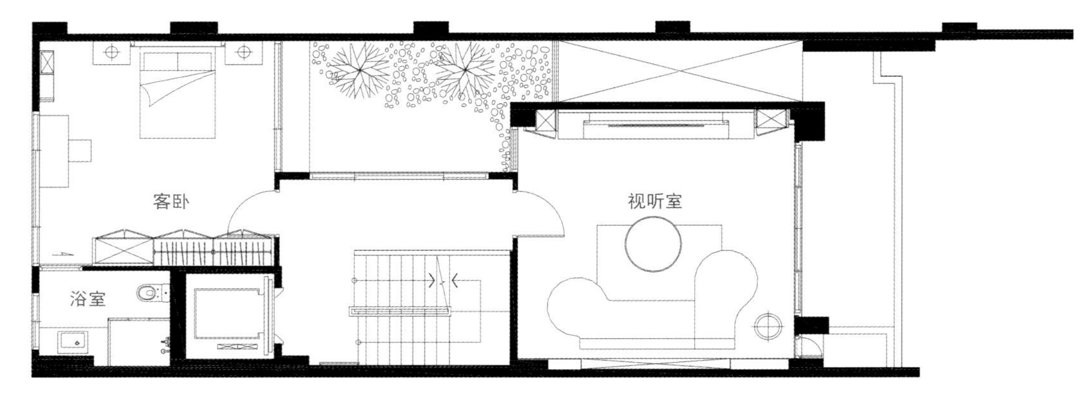
客卧
视听室
浴室

◇ NATURAL COLORS AND NATURAL LIGHTING 自然色彩、自然采光

Lights and colors determine artistic conception of the space; most color proportions are composed and stable with slightly changed basic gray tones. When entering, the item with high saturation attracts residents' eyes; perhaps it is a woolen blanket, a navy blue single chair or a piece of carpet with contrastive colors, in addition to unduplicated natural lights brought by space configuration technique, guiding residents to the natural humanistic residence which can comfort their bodies and minds with design atmosphere and to savor the space languages.

光线与色彩则以决定空间意境的角色存在，绝大多数的色彩比例是沉稳安定但却不失其中细微变化的基础灰阶色调；但当一走进，重点的高彩度单品便会吸引居者的目光，或许是一件毛呢毯被、一张藏蓝色单椅、亦或者一张高对比度色彩的地毯，再搭配运用空间配置手法导入无法复制的自然光线，其将会带领着居者进入设计氛围，细细品味接余的空间语汇，抚慰居者身心的自然人文宅邸。

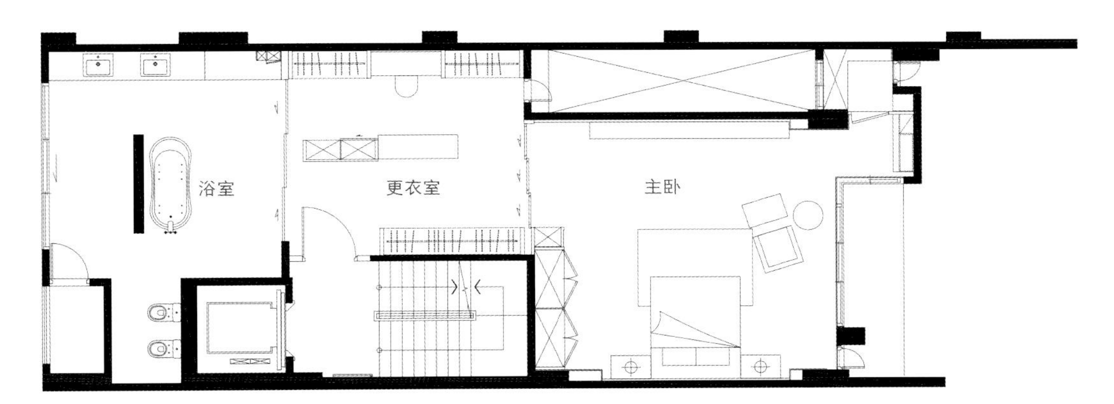

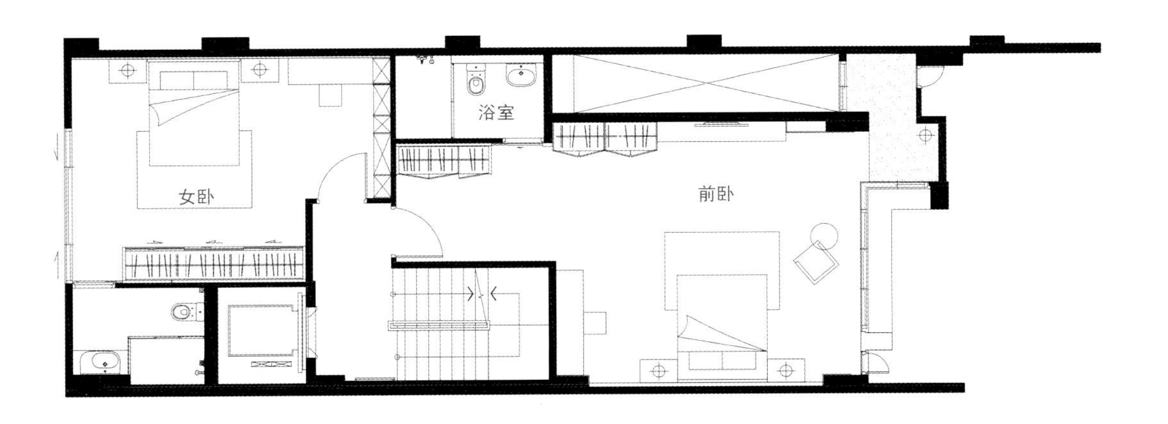

项目名称：水立方
Project name: The Water Cube

设计公司：伏见设计事业有限公司
Design company: Fujian Interior Design

设 计 师：钟晴
Designer: Qing Zhong

项目地点：台湾台北
Location: Taipei, Taiwan

项目面积：126m²
Area: 126m²

主要材料：石材、釉面砖、壁纸、木地板、杉木版等
Main materials: stone, gazed tile, wallpaper, wood floor, cedar wood board, etc.

摄 影 师：李国民
Photographer: Kuomin Lee

海辽淡水蓝·山隐绿自得

Relax Beside Nature

DESIGN CONCEPT
设计理念

This project has the whole mountain and sea scenery, large piece of windows and big balcony, which fits the good conditions of a holiday villa and features of the couple who loves traveling and sharing good taste and fun life with friends.The natural condition of this project and the lifestyle of the residents are fused together to become the design main axis of this project, namely, the holiday villa!

The designer insists the concept that "design not only touches eyes, but also touches hearts." She wants that the design is beautiful and can make the actual residents feel intimate and warm. The designer makes full use of the project itself and applies large pieces of windows to highlight the open pattern. Here, the residents can enjoy the scenery, taste their life and feel tranquility, comfort and leisure between mountains and seas!

拥有整面山景及海景、大面窗、大露台的基地空间，是符合度假别墅的优良条件，以及男女主人具有喜爱旅游、与朋友共享生活品味与乐趣的特性。此种基地的先天条件与生活者的生活形态互相融合而成为此案的设计主轴——度假别墅！

设计师秉承“设计不只打动眼睛，更能感动人心。”的理念，不只求设计要好看，更要让实际的居住者感到贴心与温暖。充分利用项目自身的优势，大面开窗，突出开阔格局，以享景品生活为宗旨，在山海之间感受一份恬静、舒适、惬意！

◇ SPACE PLANNING　空间规划

The space connected by visions contains independent spaces with different attributes. The living room is set in the center of the whole public space, making each space be used individually or together, comfortable without interference. In the open tea area and the rest platform area, the high step set in the middle can be used as chair of the tea area. The Internet, wine tasting, viewing and reading area connects with a whole set of beds in the tea area.

You are appreciating while I'm tasting tea, you are reading while I'm overlooking, we are independent yet accompany each other.

The master bedroom is divided into activity area, sleep area and sofa viewing area. The area near the door is designed as functional activity area and combines bed board with low wall, which makes the space isolated and beautiful with visual penetrable feeling. Having many window views and facing mountains and seas, the designer designs the area facing the window into a viewing area where the evening sunset can give people unlimited imagination and relaxation.

在一个视觉连结的空间内，却包含着不同属性的独立空间，将客厅设定在整个公共空间的中心点，使各个空间可以独立运用、亦可合并使用，彼此自在而不相干扰。开放式的茶艺区与卧榻平台区，中间设置的高一踏阶可以当茶艺区的座椅使用。面朝外的上网、品酒、观景阅读区，则是连接着茶艺区的一整排卧榻。

你在观赏、我在品茗，你在阅读、我在眺望，各自独立却互相陪伴。

大空间的主卧室设计也分为活动区、睡眠区，以及沙发观景区。一进门的区域设计为功能活动区，利用床头板结合矮墙的概念，亦能做出空间区隔且兼顾美观及视觉穿透感。拥有多窗景及面山面海的优良条件下，我们将面窗那一区设计为观景区，傍晚的日落能够带给人们无限的想象及放松……

◇ LIGHTING DESIGN 照明设计

Upon the original tea area and the TV cabinet, there is a crossbeam to define the space. The designer uses the "oblique tile roof type" concept of cabin, uses the "double oblique" design to make two-way lighting, which makes the slope more obvious in order to dispel the existence of the crossbeam and corresponds ups and downs to the tea area atmosphere. The choice of wood materials makes the pane of the building a single-family holiday villa.

The aroma of solid wood reconciles the cold visual sense produced by large area of white floor. The ceiling of the dining room uses imitation steel material which forms a contrast with wood to make a layering and stereo feeling. H-type steel beam groove makes lights get rid of direct and dazzling lights by the steel beam and brings out light recessivity to make them within the expected scope so as to present light and shadow, color temperature and lines of lights.

原始茶艺区及电视柜上方，拥有用来界定空间的一根大梁，我们运用小木屋的"斜瓦屋顶式"概念，使用"双斜"的设计，打上双向的灯光，能使斜度更明显以消除大梁的存在感，上下应对符合茶区氛围，选择木头的材质，将房屋的平面层塑造成独栋的度假别墅。

此外将带有香气的实木木头，调和大面积的白色地坪所产生偏冷的视觉感受，餐厅的天花板运用与木头对比性的仿钢梁材质，做出层次立体感；运用 H 型钢梁的沟槽，让打出的灯光再被钢梁消弱一层直接而刺眼的光，带出灯光隐性，使灯光限定在我们所要的范围内，呈现出光影、色温及灯光的线条。

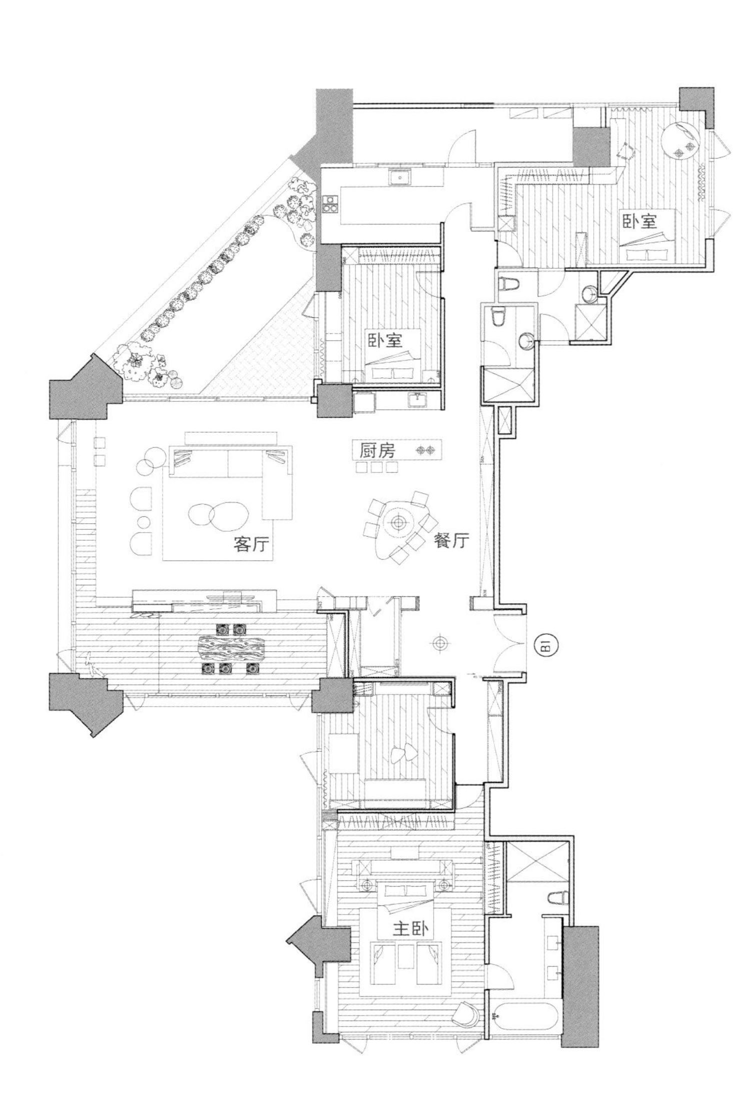

卧室
卧室
厨房
客厅
餐厅
主卧

CHANEL

◇ NATURAL COLORS 自然色彩

In addition to warm wood color and natural marble color in the whole house, the designer uses dark and light blue burnt bricks in the dining cabinet design to increase layering and vividness of the space. In order to avoid abrupt sense brought by alone existence of blue, the designer also sets up blue single chair in the living room to echo in color. In the Van Gogh color, the orange dining chairs also can make people cheerful and excited. The integral color collocation creates the wonderful atmosphere of a holiday villa.

整间除了使用温润的木头色、自然的大理石色系，我们在餐柜设计深浅两色的蓝色窑烧砖，能增添空间的层次与生动。为了避免蓝色的单独存在所产生的突兀感，我们在客厅也设置了蓝色单人椅做色彩呼应，并搭配着梵谷色系中相辅相成的橘色餐椅，亦能使人产生愉悦、兴奋的活跃感，通过色彩的整体搭配，来营造度假别墅的美妙环境氛围。

象·限 Quadrant Limitations

设计公司：近境制作
Design company: Design Apartment

设 计 师：唐忠汉
Designer: TT

项目地点：台湾新竹
Location: Hsinchu, Taiwan

项目面积：166m²
Area: 166m²

主要材料：钢刷木皮、雪白蒙卡烧面、柚木实木条、黑铁、茶镜等
Main materials: steel brush wood veneer, white monka burnt surface, teak solid wood batten, black iron, tawny glasses, etc.

DESIGN CONCEPT
设计理念

The flowing air, filtered lights, tranquil music and pleasant temperature create a poetic space. Within the dimensions of time and space, it creates intangible and tangible masses; within the quadrant of masses and men, it creates parallel interchanges; finally it initiates a deep exchange between people and space. By using light and shadow, it creates a world of "quadrant and limitation".

In this world, there are the most marvelous and wonderful Collision of texture, light, shadow and soul collision. The designer uses his own talent to endow this expressive space with more humanistic temperature. No matter you are walking or staying, you can deeply perceive its inner emotional changes.

通过流动的空气、过滤的光线、宁静的音律、宜人的温度，形塑出一个具有诗意的空间。在时间与空间的维度，创造量体虚实，在量体与人的象限，创造平行交错，终于展开一段人与空间的深度交流。用光影，创造一个"象·限"世界。

在这个世界里，有最奇妙最精彩的质感碰撞、光影碰撞以及灵魂碰撞，设计师用自己的才能赋予了这个会表达的空间更多的人文温度。无论你是行走，亦或者驻足，都可以深切的感知它内在的情感变化。

◇ DECORATIVE MATERIALS 装饰材料

The smooth texture of the steel acrylic panels is paired with rugged granite. The establishment and division of these simplistic materials emphasizes the original textures of the materials. The layered textures of the lines ingeniously meld the two elements together. Light and shadows are combined in the quadrant, becoming important elements that connect the spaces.

钢烤面板光滑质地，搭配粗糙肌理的花岗岩石，纯粹的材质设定及分割，不仅让原有的建材质性更为出色，线条的层次感更巧妙将两种元素融为一体，象限中，结合光影，形成连接场所的重要元素。

◇ SPACE PLANNING 空间规划

The foyer's screening shelves, the cabinets at the back of the sofa, the dining table island, the wall mass; the location and layout of each item, reconnects the relations between each area, recombining the horizontal and vertical planes and completing the space and the ceremony.

玄关屏柜、沙发背柜、中岛餐桌、墙面量体，各对象的座落、配置，重新链结各场所关系，重组平面象限与立面，成就整个空间和仪式。

◇ NATURAL LIGHTING 自然采光

Light is used to express the pure textures of materials. Clean linear light sources define the solid mass framework, both tangible and intangible, imbuing transparency to the overall perspective, also extending and heightening the space, which is a metaphor for the dimensional quadrants. Follow the path of light and shadow, let the ambiance of the overlapping modern culture wash over you. The TV wall forms the axis in the creation of different quadrants of space, creating a space sense of different quadrants. In this moment, people are brought closer together.

利用光，展现材质的纯粹肌理，利落的线性光源，刻画出厚实的量体框架，亦虚亦实，让整体视角更具穿透力，也提高空间与空间的延伸感，更隐喻整个象限维度。逐着光影的路径，随着现代人文交迭的氛围，以电视墙面量体为轴心，创造出不同象限的空间感，此时此刻，拉近了人与人的距离。

移步换景 兼容并蓄
Change a Scene Every Step, Embrace Inclusiveness

项目名称：如是观
Project name: Such an Idea

设计公司：白金里居空间设计
Design company: Platino Interior Design

设 计 师：林宇崴
Designer: Yuwei Lin

项目面积：225m²
Area: 225m²

主要材料：意大利进口板岩、德国进口厨具、Hunter Douglas窗饰等
Main materials: Italian imported slate, Germany imported kitchen ware, Hunter Douglas window decoration, etc.

摄 影 师：吴启民
Photographer: Qimin Wu

DESIGN CONCEPT
设计理念

“Everything in the world is as fantasy and false as the morning dew and lighting with moments of birth and death. We should hold such an idea.” The space is tangible while the value of creation is intangible. They agitate and intersect and develop freely, following the shape, essence and light of the space at will and not persisting in creating new words.

The couple who lived in France has Western background and Eastern culture. They choose a lot of imported furniture with concise lines, collect many Zen-like art works and want to display them in the house. The designer goes with the thought to make the best use of these Chinese antique collections so as to successfully reflect the taste of the owner and to create cultural flavor and texture of an art city.

“一切有为法，如梦幻泡影，如露亦如电，应作如是观。”空间是有形的，创作所赋予的价值是无形的，两者激荡交会，随心发展，随兴所至，顺着空间的形态、本质、光线，不执着为赋新词。

曾旅居法国的屋主夫妇，兼具西方背景和东方文化，在大量使用线条简洁的进口家具的同时，亦收藏许多具有禅意的艺术品，希望可以将收藏一一陈列于居家空间。白金里居空间顺势而为，让这些中式古董藏品物尽其用，成功彰显屋主品味的同时，营造如艺术之都的文化气息与质感。

VIEW

VIEW
DESIGN
HOTEL
YEAR
BOOK

◇ DECORATIVE MATERIALS 装饰材料

In porch, dark color wood veneer is used to decorate cylinder, collocating with embedded gray lens to hide the function of shoe cabinet. Concave display space with peaceful statue of art can deposit body and mind. The dining and kitchen area uses gray laminate hard plastic sheets to endow natural texture with the elevation imitation stone. The well-chosen wall clock adds variability to the space. At the same time, there are many reflective glossy materials in the whole space, reflecting the exquisite inverted image of the space so as to increase the width and depth.

在玄关，以深色木皮修饰柱体，搭配镶嵌灰镜质材，将鞋柜机能隐藏。内凹展示空间则摆放祥和的神像艺术品，具有沉淀身心的作用。餐厨区则采用灰色系美耐板材质，赋予立面仿石材的天然肌理；搭配精心挑选的时钟墙，增添空间变化性。同时整体空间中多用具反射特性的亮面质材，折射出精致的空间倒影，借此增加纵宽和深度。

◇ SPACE PLANNING　空间规划

In the preliminary roughcast stage, the designer follows lights and the width of life to plan the space. The public space uses open layout to plan the study, living room, dining room and kitchen. The dining and kitchen area follows the hostess's preferences and needs and adopts the island kitchen design with Germany imported kitchen ware. The surroundings are broad walkways to form the circulatory and fluent life layout so as to promote family interactions.

初期的毛胚状态，设计师依循光线和生活的宽阔度规划出空间，公共空间以开放动线规划出书房、客厅、餐厅和厨房。餐厨空间则顺应女主人喜好和需求，以德国进口厨具中岛式厨房设计，四周让出宽阔走道，形成循环流畅的生活动线，增进家人的互动性。

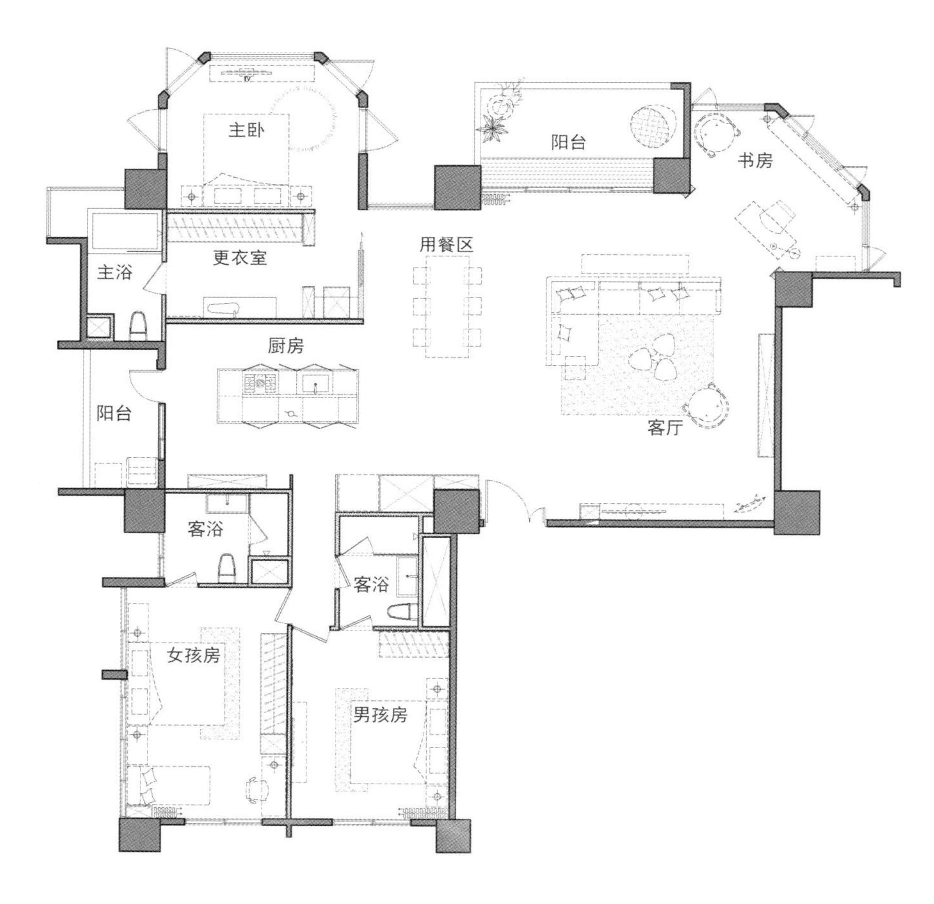

◇ NATURAL COLORS　自然色彩

It abandons the technique of gorgeousness and jumping color of a general mansion, uses calmness to present extraordinary manner and carries on gradation, temperature and saturation of color to highlight the details. The basic tone of earth color presents sedate designs, which not only highlights the collections but also foils the extraordinary texture of imported high-quality furniture.

舍弃一般豪邸华丽、跳色的手法，以沉稳表现不凡的气度，并就色阶、色温、色相连贯相承，让细节更为突出。透过大地色的基础色调呈现沉稳的设计，不仅让收藏品更为突出，也衬托出进口精品家具的不凡质感。

项目名称：木艺术静品宅
Project name: Wood Art Tranquil Residence

设计公司：尚邑国际室内装修有限公司
Design company: KENNY & C INTERIOR DESIGN

设 计 师：吴志勇
Designer: Kenny Wu

项目地点：台湾桃园
Location: Taoyuan, Taiwan

项目面积：228m²
Area: 228m²

主要材料：柚木、栓木、铁件等
Main materials: teak, ash, iron, etc.

木质精品 人文艺术

Wood Boutique, Humanistic Art

DESIGN CONCEPT
设计理念

Sir Reynolds, a famous British painter in the 18th century, once said: “Art makes nature more perfect”, while the inspiration of art is mostly from nature.The delicate wood condo is made of two natural wood veneers, namely Teak & Ash, is embedded with the soul of earth and abandons superfluous and flatulent decorations. The design full of strong individual style can be clear seen in the living room and reveals the most authentic interaction between human and nature, warm and luxuriant.

This exquisite timber space is made of wood elements, combines with outdoor courtyards, bringing endless green scenery into interior and making the indoor and outdoor naturally into a whole. The warm breath of wood and vital green of courtyard supplement each other. The simplified interior design abandons exaggerated furnishings and uses wood and natural green to bring the owner the simplest and easiest, casual and comfortable life .

18 世纪英国名画家雷诺兹爵士曾言：“艺术使自然更完美”，而艺术的灵感来源大多为自然环境。此案以柚木与栓木两者结合的内装，将大地之母的灵魂镶嵌其中，摒除多余且浮夸的装饰。充满浓厚个人风格的设计在客厅便可一目了然，倾吐人类与自然最真实的互动、温暖而丰润。

整体以木原素构成精致的木空间，结合室外庭院，将无限的绿意景致引入室内，使得室内外自然天成，浑然一体。木质温润如玉的气息和庭院生机的绿意相辅相成，删繁就简的室内设计，摒弃夸张的饰品，以木质和自然绿意带给屋主最简单、轻松又不失自由的惬意生活。

◇ DECORATIVE MATERIALS AND SPACE PLANNING 装饰材料、空间规划

The herringbone wood beam near the French window holds up, as if using tree images to symbolize the owner's solid and forceful will. The carved characters on the wall add nifty and clever fun, whispering quietly. Backwards and entering into multi-fuctional space for dining, reading and interacting, the designer quietly brings droplight in industrial style into the space, adding a conflict beauty. The scattered bookcases on the wall jump out of the role of storage and become the furnishings.

The opposite wall and seats are based on wood to foil the colors, composed and interesting. The irregular fine wood grilling and Winston glass write about composite material languages, at the same time echo the power and beauty of the herringbone wood beam. The bedroom has its own style; it is based on creamy white tone and matches large area of floor window with gauze curtain, making the lights soft and warm.

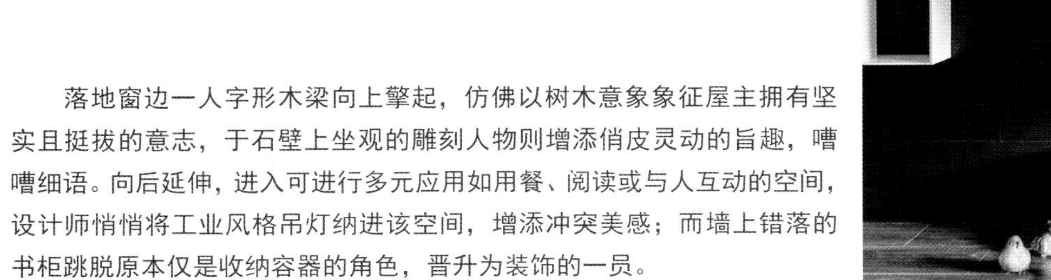

落地窗边一人字形木梁向上擎起，仿佛以树木意象象征屋主拥有坚实且挺拔的意志，于石壁上坐观的雕刻人物则增添俏皮灵动的旨趣，嘈嘈细语。向后延伸，进入可进行多元应用如用餐、阅读或与人互动的空间，设计师悄悄将工业风格吊灯纳进该空间，增添冲突美感；而墙上错落的书柜跳脱原本仅是收纳容器的角色，晋升为装饰的一员。

对面墙壁与座椅皆以木质为基底衬托缤纷色彩，沉稳中不失童趣，一旁不规则细木格栅为云丝玻璃书写关于复合材质的语言，同时呼应人形木梁的力与美。卧室则自成一格，以米色白色调为基底，大面积落地窗搭配纱质帘幕，使光线柔和而不失温度。

ddn
MILANO
DESIGN
SELECTION 2013

◇ BRINGING SCENERY INTO HOUSE 引景入室

As for the outdoor spacial garden, the designer matches grass and wood with nifty accessories, is good at using lights, brings scenery into the room and creates a dreamy and fresh atmosphere at night. The multi-functional rest space is bright and spacious and combines with child interest and calmness, giving the resident high security and practicability and elegant and exquisite life fun. Creating a space for dealing with oneself should be loved by modern people and live up to the humanistic and artistic design space.

针对户外大空间庭园，更以草木为对话，搭配俏皮配件并善用光线，邀景入室，并在夜晚营造梦幻而清新的氛围。多元休憩空间的规划明亮且宽敞，同时结合童趣与沉稳，给予使用者高度安全感以及实用性，却不失优雅与精致的生活旨趣。无论高谈阔论、品茗小憩亦或是冥想沉淀，创造自己与自己的相处空间，应深得忙碌现代人喜爱，不负人文与艺术结合的设计空间。

对话空间 Dialogue Space

项目名称：淡水卢邸
Project name: Tamsui Lu Residence

设计公司：隐巷设计顾问有限公司
Design company: XYI Design

设 计 师：黄士华、孟羿彣、袁筱媛
Designers: Mac Huang, Carrie Meng, Eva Yuan

参与设计：彭芝榕
Cooperating designer: Lousie Peng

项目地点：台湾台北
Location: Taipei, Taiwan

项目面积：231m²
Area: 231m²

主要材料：盘多磨、胡桃实木、超薄瓷板、灰镜、黑纤、银狐、赛利石、不锈钢、黑烤漆玻璃、特殊涂料漆等
Main materials: Pandomo, walnut solid wood, super thin porcelain plate, gray glass, black fabric, silver fox marble, Salina stone, stainless steel, black paint glass, special paint, etc.

DESIGN CONCEPT
设计理念

How do elders and married young generation live together and retain their own privacy?

The owners are two elder and younger couples. The young hostess who lives abroad for years hopes to have a personalized rest space. The elder's life focus is at home; and friends and relatives often visit them to party; they want to have a comfortable and relaxing space.

There is no framework for dialogue. We use the whole concept of "home" to contain independent living spaces for two generations and expect to have more dialogues between them, then gradually come up with the idea of "frameless dialogue". We get through the two houses, break the original pattern and dismantle the seven small rooms to construct three independent life places including the public space, bedroom for the elder couple and bedroom for the young couple. By breaking the "framework" of the existing pattern, we create an more open and flexible "dialogue" space.

长辈与成家的年轻世代如何共同生活又保有各自的隐私空间呢？

屋主为两对年长及年轻的夫妻。年轻的女屋主长年旅居国外，希望拥有较个性的休憩空间，长者的生活重心在家，常有亲友来访与聚会活动，希望空间能舒适令人放松。

没有框架的对话。我们以一个“家”的整体概念包容两代人各自独立的生活空间，期许有更多的对话在这两代人之间发生，于是逐渐萌生了“无框对话”的想法。将两户打通并打破原本的格局，将七个狭小的房间拆解，重新建构成三个独立机能的生活场所，其中包含公共空间、年长夫妻卧室、年轻夫妻卧室。借由打破既有的格局“框架”，萌生出更宽阔更灵活的“对话”空间。

◇ DECORATIVE MATERIALS 装饰材料

Sun lights enter into the public space constituted by the living room and dining room from the balcony. The warmth of sun, the fragrance of wood, the softness of fabric and the calmness of stone create a comfortable and bright rest area, which makes people want to stay inside. The dining room uses the industrial style wine shelves and dining table as the visual focus, conveying the hospitality of the host. Log aromas mixes with subtle rough flavor. The fusion of two kinds of design style also is the transiting gateway to the young couple's private space, that is design infiltrated from everyday life.

日光从阳台洒进由客厅与餐厅构成的公共空间，阳光的温度、木头的香氛、纤维的柔软、石材的沉稳形成舒适明亮的休憩地带，将人驻留在此。餐厅以工业风格的酒架及餐桌作为视觉中心，传达出屋主热情的待客之道。原木的芳香混合着细微的粗犷气息，两种设计风格的融合之处亦是通往年轻夫妻私密空间的转折场所，即从日常间渗透的设计。

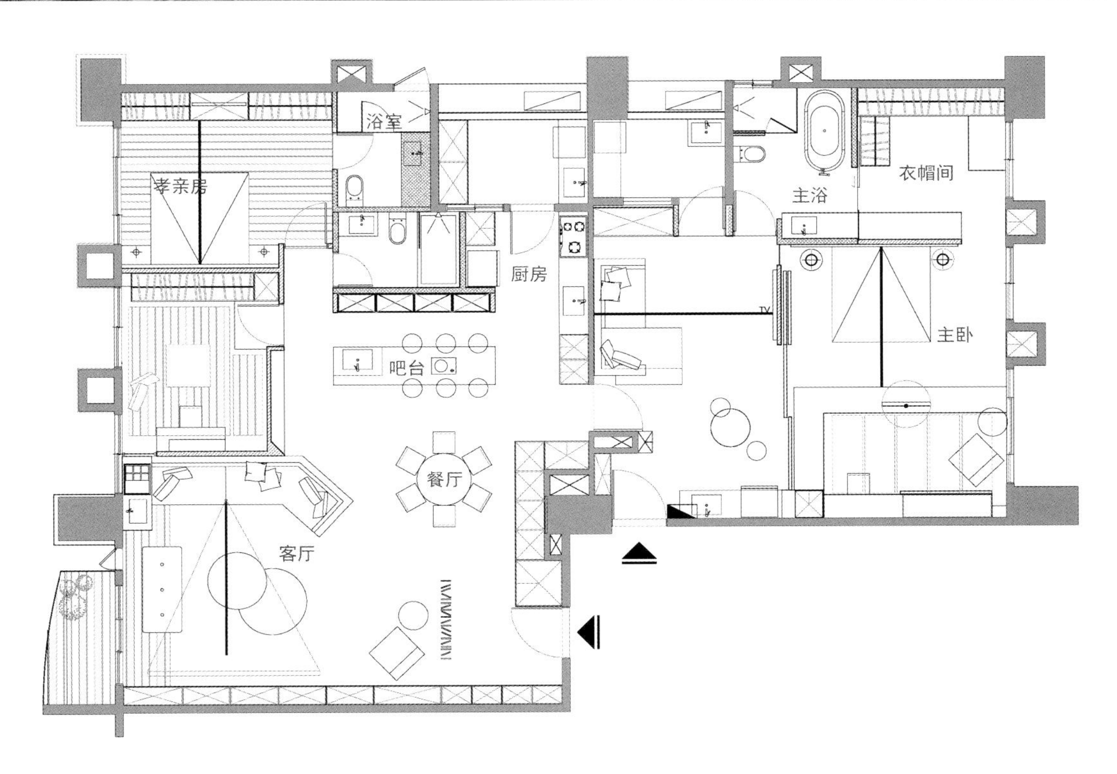
浴室
孝亲房
衣帽间
主浴
厨房
主卧
吧台
餐厅
客厅

◇ NATURAL COLORS 自然色彩

The elder's room hides a lot of wood color storage cabinets, which accords with the user's life habits. The young couple's room gives priority to personalized industrial vocabularies and uses a lot of warm color logs and monochromatic bright soft decorations to melt the cold bold lines and to make here a relaxing and comfortable living space with distinct style.

长辈的房间隐藏着大量木色的收纳柜体，符合使用者的生活习惯。年轻夫妻的房间以个性鲜明的工业语汇为主体，运用大量暖色的原木以及单色明亮的软饰，融化冰冷的粗犷线条，使这里成为风格鲜明、放松惬意的居家场所。

心生向往 幸福同堂
Heart Yearning, Happiness Together

项目名称：八德洪宅
Project name: Bade Hong Residence

设计公司：怀生国际设计有限公司
Design company: WYSON DESIGN

设 计 师：翁嘉鸿
Designer: Wyson

项目地点：台湾桃园
Location: Taoyuan, Taiwan

项目面积：396m^2
Area: 396m^2

主要材料：木皮、美耐板、烤漆、石纹元素、铁件、玻璃等
Main materials: wood veneer, laminate hard plastic sheet, stoving varnish, marbling element, iron, glass, etc.

摄 影 师：墨田工作室吴启民
Photographer: Moooten Studio

DESIGN CONCEPT
设计理念

This project is a house for a happy family with three generations, which has six floors with nearly 400 square meters. The couple is engaged in network marketing and influenced by creative life so that they love the distinct design feature which is displayed in the household style. In addition to bright lighting and appropriate furniture, it creates a comfortable and clean indoor environment.

The couple who loves life wants to create a comfortable home, and the dialogue between spaces and the flow of light and shadow convey the designer's ideas. The independent space of every floor differentiates work from life and separates public area from private area to make life free and non-interfering among families and to keep the happy time together.

本案是一个三代同堂的幸福之家居所，有着六层近400平方米的住宅面积。屋主夫妻皆从事网络营销，耳濡目染创意生活，造就热爱不俗的设计特色，这样的特质也流露在居家风格里，再搭佐明亮的采光，适宜的家具，便能营造出舒心洁净的室内氛围。

热爱生活的业主将这里打造成舒适之家，空间的对话和光影之间的流转传达出设计师的意念。每层独立的空间不仅将工作与生活区别开来，更将公领域和私领域分隔开，让彼此之间的生活更加自由不相干扰，又留住了一家人在一起的欢乐时光。

◇ SPACE PLANNING 空间规划

Located in Bade readjustment area with broad surroundings and good lighting, this project has a pattern of every floor a different space. The first floor is the office where the couple can be engaged in network marketing. The second floor is public space with living room, dining room and kitchen. The third floor is master bedroom and master bathroom with perfect supporting configurations to provide the couple a better life quality. The fourth floor is two rooms for kids. The fifth floor is the elder's room. The sixth floor is the landscape garden and the place for dolls. In the process of design, there is only difficulties in planning the air-conditions. Because the beams are in demarcation points of different spaces and the designer doesn't want to pull down the ceiling to cover the lines. But finally the problem is overcame by modeling.

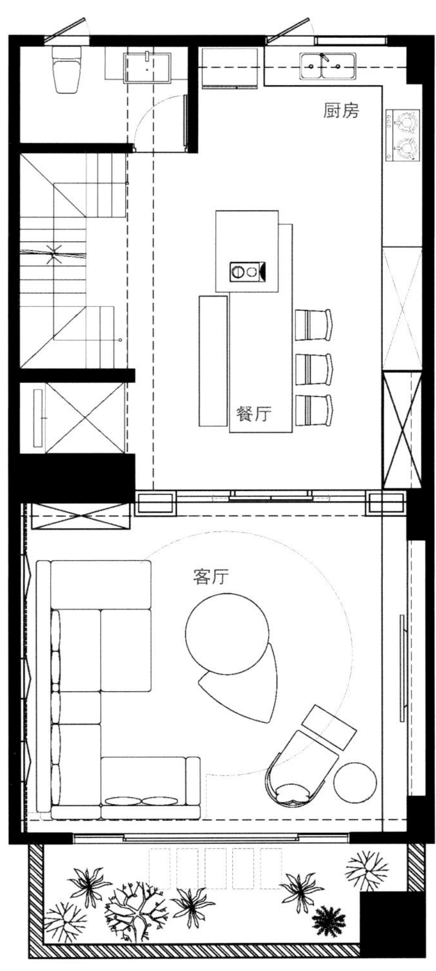

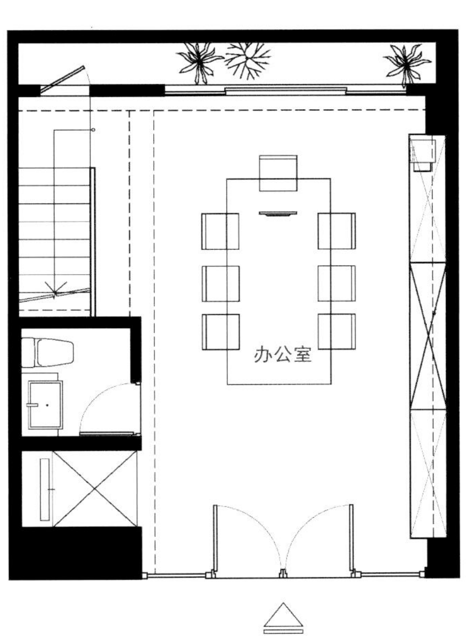

本案位于八德重划区，周围宽广，光线非常良好，格局为一层楼一种空间，一楼是办公室，是从事网络营销业主夫妻办公的场所。二楼是公共空间，包含客厅与餐厅厨房。三楼整层是主卧室和主浴室，完善的配套设置为业主提供更好的生活品质。四楼是两间小孩房，五楼为孝亲房，六楼则是公仔放置区及景观花园。设计的过程中唯有空调的规划较为困难，因为梁柱位于不同空间的分界点，但我们又不想拉低天花板去包覆管线，最后还是以造型克服了这个问题。

◇ DECORATIVE MATERIALS 装饰材料

In public areas of the second floor, the tea table is made of branches with annual ring lines in the living room; the sofa background wall is created by straight and oblique wood grains while the TV wall and the ceiling are outlined by black lines. The nearby display shelf continues the change of lines. Concise colors and neat texture materials make the vision of interior space more vivid and changeable. The designer uses modern materials such as wood veneer and laminate hard plastic sheet to pave concise texture. The changes of vertical lines create a three-dimensional and layering extraordinary space.

二楼公领域枝干年轮的线条化身客厅的茶几，沙发背墙以直、斜线条交错的木纹接构而成，电视主墙到天花板皆以黑色调的线条勾勒，一旁的展示层架同样延续线条的变化。简约用色与利落感材质打造的室内空间，视觉更显生动与变化，运用木皮、美耐板与石材等现代化质材，铺陈出简约质感，在透过立面线条的变化，刻划出立体、渐层视感的非凡空间。

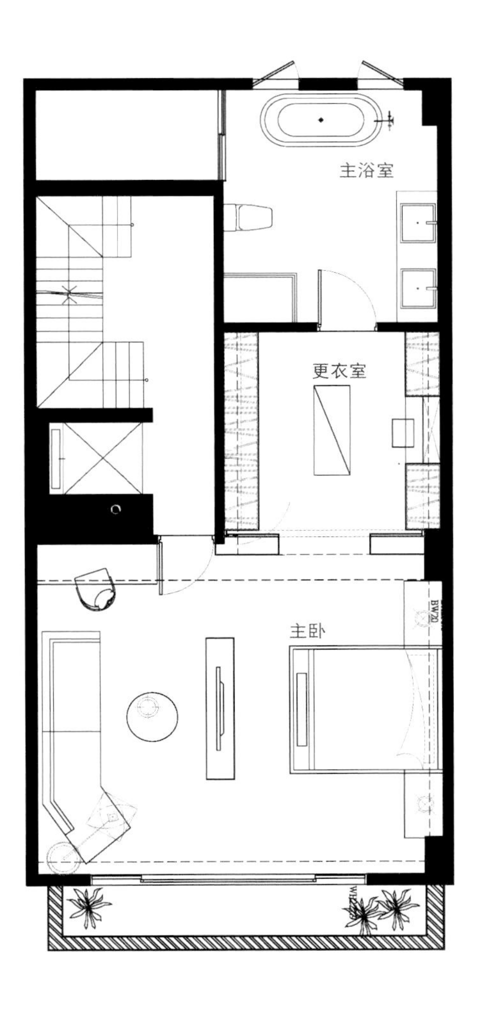
主浴室
更衣室
主卧

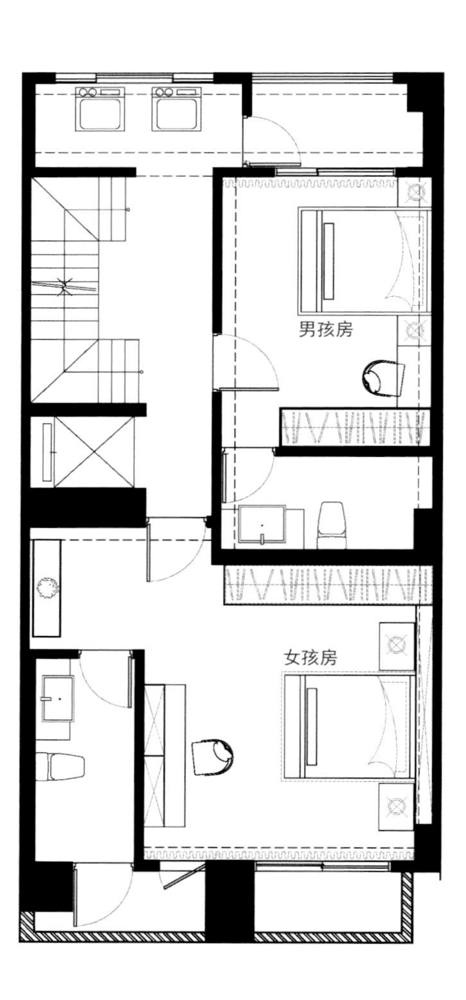
男孩房
女孩房

◇ NATURAL LIGHTING　自然采光

The designer originally creates space areas and skillfully introduces natural light to endow the interior with plentiful light and shadow. Broad and spacious French windows in every area bring in outdoor lights and glass sliding doors divide the living area and dining area. Open balcony in every floor is transparent and bright, enhancing the communications between indoor and outdoor spaces and continuing outside natural and bright lights. Open and flowing space breath makes the whole interior more extraordinary.

设计师匠心打造区域空间，良好地引入自然光线使得室内光影充足，各区域宽大朗阔的落地窗导入室外光线，同时以玻璃拉门界定客、餐领域，每层开阔的阳台通透爽朗，加强室内外之间的空间对话，延续户外自然透亮的光源，开放、流动的空间气息，让整体室内气度非凡。

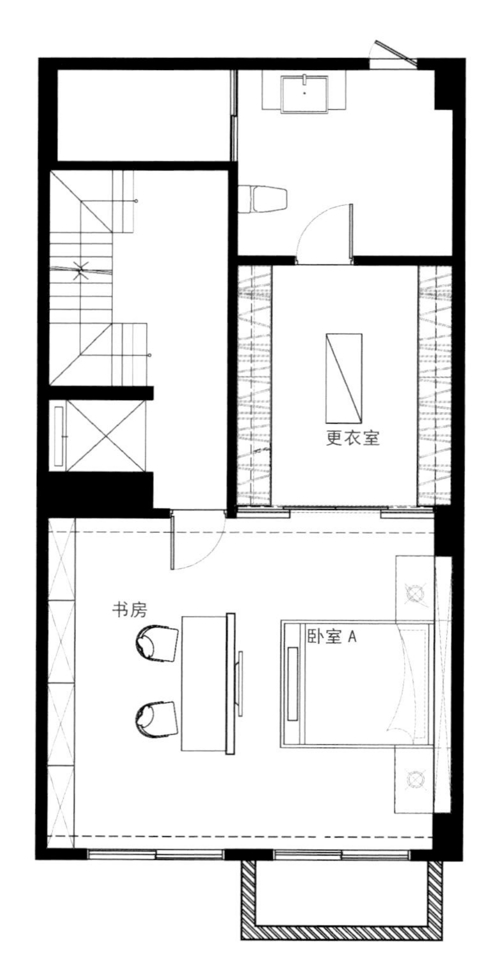

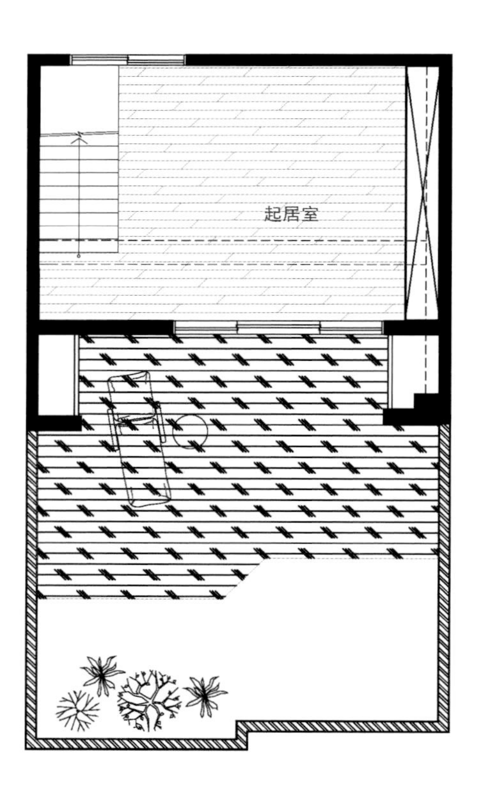

驻留空间纹理节奏

Staying with the Space Grain Rhythm

项目名称：林口张宅
Project name: Linkou Zhang Residence

设计公司：怀生国际设计有限公司
Design company: WYSON DESIGN

设 计 师：翁嘉鸿
Designer: Wyson

项目地点：台湾新竹
Location: Hsinchu, Taiwan

项目面积：300m²
Area: 300m²

主要材料：铁件、木皮、清水模特殊漆、石材、珪藻土涂装等
Main materials: iron, wood veneer, clear water mould special paint, stone, kieselguhr coating, etc.

摄 影 师：朴叙空间创意有限公司
Photographer: MD PURSUIT

DESIGN CONCEPT
设计理念

Material grains can endow the dialogue between space and people with authenticity and purity. The guide switch of light and space material texture is the design thought which the designer tries to convey through this project. How to make every space become the focus of the residents and the intuition for them to stay through light guidance and space transformation of material grains is the unique design concept and perform technique of this project. Clear material grains make life more profound so that the space memory naturally extends to the ups and downs of material grains.

材质纹理能让空间与人的对话具有真实性与纯粹性，光线与空间材质纹理的引导切换，是设计师试图在此案传达的设计理念。如何使各空间于收放之间，借由光线引导与材质纹理的空间转换，达到居者空间聚焦以及驻留身体性直觉，是本案独具特色之设计概念与执行手法。清晰的材质纹理让空间生活变得更深刻，使得空间记忆自然延伸于材质肌理的起伏之间。

◇ NATURAL LIGHTING 自然采光

White tone of the core space extends to the dining area; horizontal white-washed wood grain and vertical wall cabinet echo with each other; the light vertical and horizontal corresponding relationship creates the space atmosphere and naturally interprets the space stability of warn wood furniture. The residents stay with space life memories with families between lightness and gentleness and daylight joys.

由核心空间白色基调横向延伸至生活用餐区域，并以水平刷白木纹以及对向垂直错落墙柜彼此呼应，以垂直水平之轻淡对应关系围塑空间氛围，自然诠释温润木质家具之空间稳定性。使居者在轻淡温和与日光写意之间，驻留刻写与家人之空间生活记忆。

◇ BRINGING SCENERY INTO HOUSE 引景入室

The master bedroom features a large area of lighting and deliberately keeps a horizontal rest area near the window to give it a comfortable and quiet atmosphere, to make residents enjoy the peace of communicating with outside urban space and to bring in natural scenery. Wood line grain and vivid splicing technique transfer the horizontal sense near the window to floor of main bed area. The black and white contrast deepens the core sense and stability of main bed area to make it a period of light scenery and space grains.

主卧房特色为大面积采光，并刻意于窗边置入水平驻留区域，让主卧房能在舒适安静的氛围中，使居者享受与外在都市空间对话的宁静感和引入空间中的自然景色。借由木质线条纹理，附加生动拼板手法，将窗边的水平感转折落地于主床区域。并以黑白对比调性，加深主床区域的核心感与稳定感，成为光线景色与空间纹理连动之间的句点。

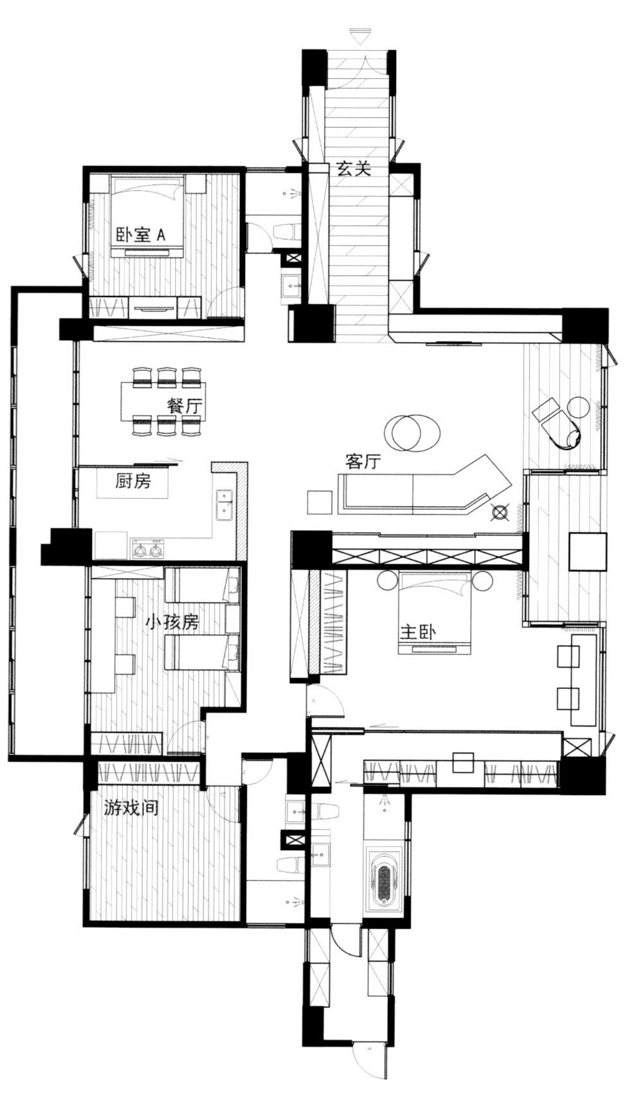
玄关
卧室 A
餐厅
客厅
厨房
小孩房
主卧
游戏间

◇ NATURAL COLORS 自然色彩

The pattern of master bathroom is clear and decent; the open horizontal lighting makes the space lines corresponding and clear. The gray space uses natural and vertical rise and fall operating technique to directly definite functions of every space and the echo of stays. The conversion of each functional area exquisitely uses different geometric brick grain texture between light and dark tones to depict the readability of space stays. The residents can be bathed in the symphony of space in the combinations of lights, ups and downs and texture orientation.

主浴室格局清率大方，开阔的水平采光使得空间线条对应清晰。灰色空间基调主以自然水平起伏之操作手法，直率定义各种空间功能与停留的呼应性。各子功能区域的率直转换之间，更在同色调深浅之间细腻地以不同几何砖纹肌理，刻画空间停留之可读性。使居者能在光线、水平起伏与肌理定位的组合之中，浸浴于空间凝冻之交响曲。

商务饭店风 精品时尚宅

Business Hotel Style, Exquisite Fashionable Residence

设计公司：千彩胤空间设计有限公司
Design company: Cty Space

设 计 师：李千惠
Designer: Peggie Lee

项目地点：台湾高雄
Location: Kaohsiung, Taiwan

项目面积：265m²
Area: 265m²

主要材料：木皮、石材、铁件、薄砖、特殊漆、壁纸等
Main materials: wood veneer, stone, iron, thin tile, special paint, wallpaper, etc.

摄 影 师：刘欣业
Photographer: Xinye Liu

DESIGN CONCEPT
设计理念

As a residence for the mainland owner to sleep occasionally in Taiwan, this project is a rest house as well as a staff dormitory. Because the decoration budget is limited, the owner wants the whole style simple and decent. Considering that the owner is old, the designer changes the orientation of the stair and decreases its slope to reduce the risks when walking up and down stairs. The whole house abandons complicated modeling changes. Concise lines and special grains of materials and piles of rich layering expressions endow the space with high texture atmosphere.

Orientated in business demand, every bedroom adopts suite configuration with well-equipped life functions including independent bathroom and basic collecting cabinet. There are also USB sockets in the bedside. All imitates the hotel mode to create the most comfortable and convenient rest spaces. In addition, there is also Wi-Fi base station interior, providing hassle-free network function. So the owner or stuff who will spend the night here can surf the internet and send and receive mails at any time.

本案作为陆商屋主偶尔来台过夜之所，同时也是招待所和员工宿舍。因装修预算有限，希望整体风格以简单、大方为主，并考虑屋主年事已高，在客变时即更改楼梯坐向、降低坡度，减少上下楼时的危险性。全室没有采用过于繁复的造型变化，仅透过简约线条与材质特殊纹路，堆栈丰富的层次表情，赋予空间高质感氛围。

以商务诉求为导向，每间卧室皆采用套房配置，具备独立卫浴、基本收纳柜等齐备的生活机能，床头亦设有USB插槽，效仿饭店模式给予最舒适、便利的休憩空间。此外，室内也装设Wi-Fi基地台，提供无死角的网络功能，让前来过夜的屋主或员工，可随时上网、收发信件。

◇ DECORATIVE MATERIALS 装饰材料

The porch chooses rust thin tile to connect the shoe cabinet with door piece of the storeroom. The mottled old texture and the iron frame shelf form a new and old contrastive beauty. Marbles are covered in the TV wall, which becomes the visual focus of the open field. Some parts retain the display area which can be used to put electrical equipments and furnishings and avoids to be heavy on the whole. Walls near the dining table are paved with stones, collocated with layer boards and self-assembled red wine shelf, which foils the temperament of the entire space.

玄关选用锈面薄砖，串联鞋柜与储藏室门片，其斑驳仿旧质感与一旁的铁件展示架，形成新旧融合的对比美感。大理石作为电视主墙，成为开阔场域内的视觉重心，局部保留展示区域，可用以放置电器设备或装饰品，也避免整体过于厚重。餐桌旁的壁面以石材铺陈，搭配层板与自行组装的红酒架，烘托整体空间质感。

◇ SPACE PLANNING 空间规划

Because the original stair is too sharp, the designer changes the stair orientation and slope to ensure the convenience and security of the owner when walking up and down stairs. In the dining room, the ceiling is covered with wood boards to form clear field boundary and to inject the space with some warm textures. Designs of the open public area retain the original kitchen equipments to make the dining and kitchen area become the perfect social place to interact.

因原本楼梯过陡，设计师于客变时更改楼梯坐向与坡度，以确保屋主在上下楼梯时的方便性与安全性。餐厅区天花铺贴木板，形塑明确的场域界定，亦为空间挹注些许温润质地。开放式公领域设计，保留建商原本的厨具设备，让餐厨区成为互动性绝佳的社交场所。

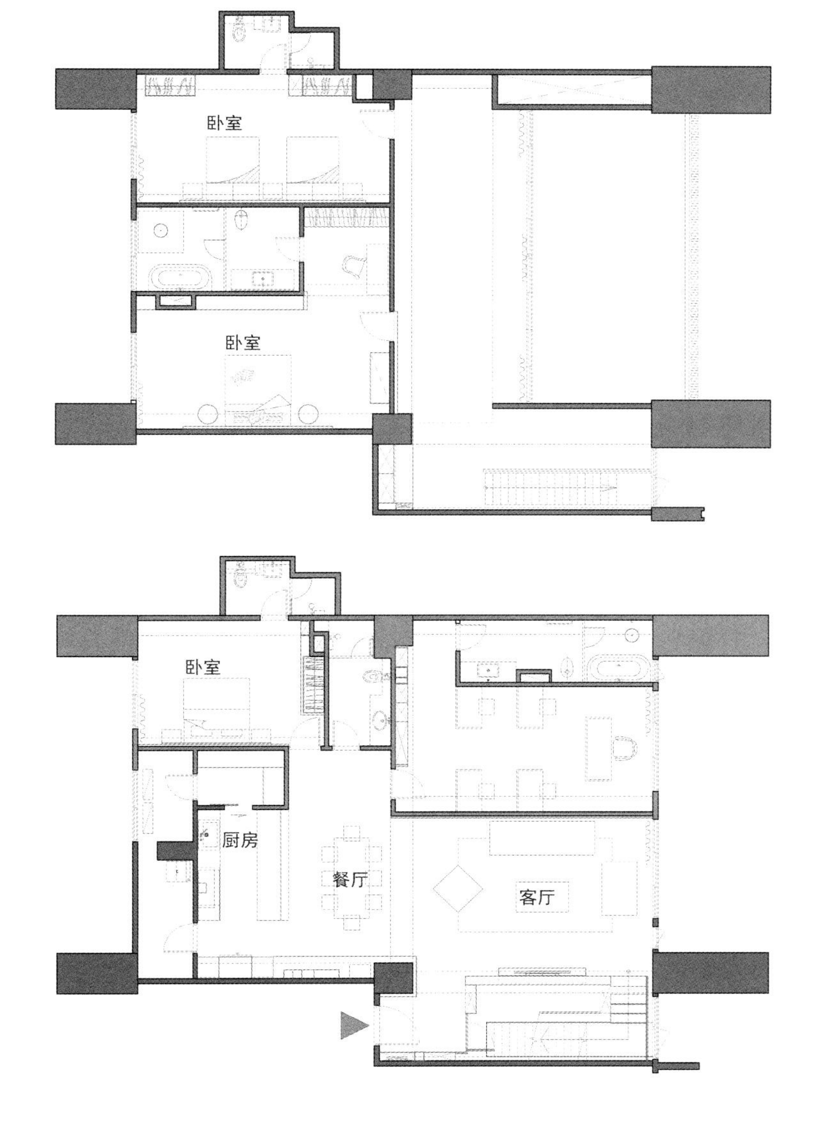

◇ NATURAL COLORS 自然色彩

The bedside uses coffee stretched fabric, collocated with well-organized lines, which injects taste and texture into the space. According to the owner's preference, the whole TV wall replaces common wood veneer with cream color paint glass, embedded with iron display cabinet, adding visual layering of the elevation. The subaltern room is based on warm wood tone, creating enjoyable and leisure rest atmosphere.

床头采用咖啡色绷布铺陈，搭配疏密排列的线条变化，为空间注入品味与质感。电视主墙因应屋主的喜好，整面以奶油色烤漆玻璃取代常见的木皮，局部嵌入铁件展示柜，增添立面的视觉层次。次卧室以温润的木质色调铺陈场域基底，营造写意休闲的休憩氛围。

质感漫入 现代品味宅

Modern Tasty Residence Full of Texture

设计公司：千彩胤空间设计有限公司
Design company: Cty Space

设 计 师：李千惠
Designer: Peggie Lee

项目地点：台湾台南
Location: Tainan, Taiwan

项目面积：139m²
Area: 139m²

主要材料：天然木皮、铁件、天然石材、锦绣、烤漆、特殊漆、玻璃、可乐石、实木桌板、百叶窗、裱布等
Main materials: natural wood veneer, iron, natural stone, splendid embroidery, stoving varnish, special paint, glass, Coke stone, solid wood table board, shatter, fabric, etc.

摄 影 师：刘欣业
Photographer: Xinye Liu

DESIGN CONCEPT
设计理念

The owner is very fond of the design style of Cty Space. So when buying a new house, the designer is invited to do the plenipotentiary plan. The hospitable owner who often invites friends to gather in the home, needs a capacious and comfortable activity area and wants to have an enlarged visual effect in the given space. At the same time it meets practical demands of storage and convenient cleanness.

The whole space is based on white and wood color, defusing into interior field from the porch. The wood ceiling is covered extensively, connects with the dust area, dining room and kitchen and hides the air conditioning at its convenience to promote design layering and texture of the space. The reflection of bright mirror makes vision extend infinitely, creating a bright and magnificent field manner. In addition to define the public area by open design, the dining room is specially placed with a big table which can accommodate many people and combines with island bar design, increasing the interaction between people when making light food here.

屋主十分喜爱千彩胤空间设计的设计风格，因此购入新屋时便找来李设计师全权规划。好客的屋主经常邀请亲朋好友来家中聚会，需要一个宽敞舒适的活动区域，并希望在既定空间中，能够创造加倍放大的视觉效果，且同时满足大量收纳和清洁便利等实用需求。

整体以白色与木色为基底，从玄关处开始漫入室内场域。木质天花采大面积铺陈，串联起落尘区、餐厅及厨房，并顺势将空调隐藏其中，提升空间的设计层次与质感。更借由明镜反射，让视觉达到无尽延伸的效果，创造明朗、大气的场域气势。除了以开放式设计界定公领域外，餐厅区也特别摆设大餐桌，可同时容纳多人用餐，并结合中岛吧台设计，能在此料理轻食，增加人与人之间的互动关系。

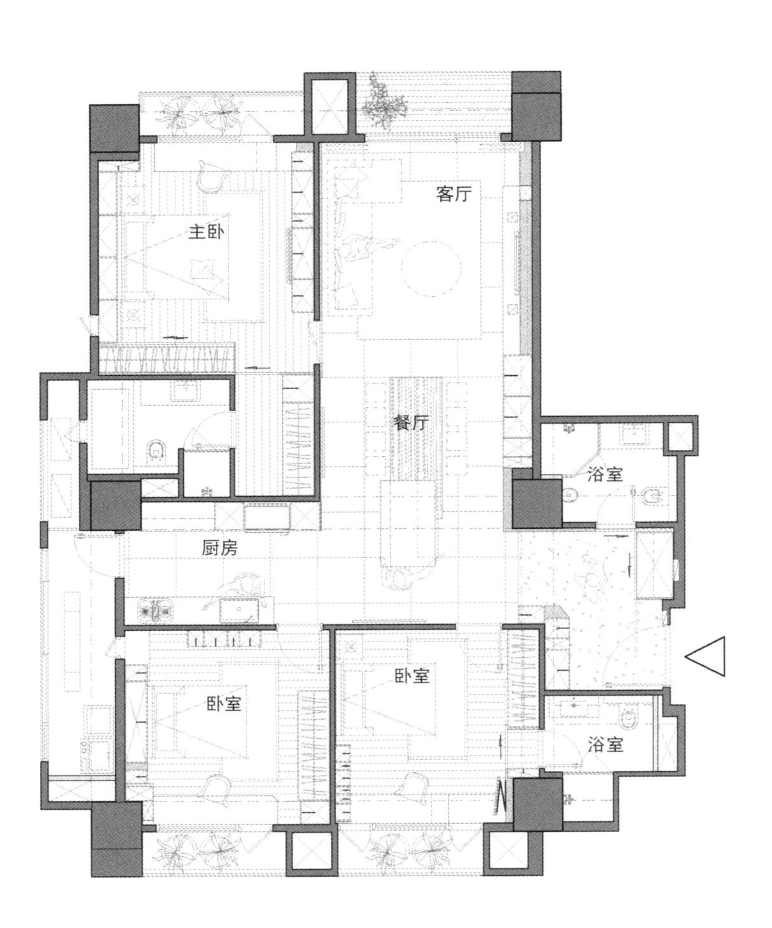
客厅
主卧
餐厅
浴室
厨房
卧室
卧室
浴室

◇ SPACE PLANNING 空间规划

It is a single-storey residence. In addition to circulation planning and layout configuration, when designing it, the owner stresses the function of storage and wants to enlarge the space. The designer makes full use of mirrors to hide a lot of storage space; the reflection effect of mirror extends the space sense and presents a modern concise and fashionable texture! The porch is placed with a cabinet; open up the mirror door piece, there is space to put clothes; close it, it can be used as dressing mirror and extends the space sense of the porch.

单层住宅，除了动线规划和格局配置外，在装修设计时，屋主也十分注重收纳的机能性，并希望能让空间更加放大。设计师充分利用镜面隐藏了大量的收纳空间，且透过镜面的映射效果，不但延伸了空间感，也表现出现代简约的时尚质感！同时玄关处便安排了丰富的柜体收纳，拉开另一侧的镜面门片，内部更蕴含置放衣物的空间，当将门片收起，平时也可作为穿衣镜使用，更延伸了玄关处的空间感。

◇ NATURAL COLORS 自然色彩

The public space links functions of the living room and dining room, paved with warm wood color and collocated with light color walls and travertine, creating a rich contrast of materials and colors. The master bedroom uses concise and sedate tone to deposit the atmosphere. It uses white shatters near the window to enable the owner to adjust the interior light freely according to read and rest requirements and to feel comfortable and relaxing in the bedroom. Except for general cloth storage space, the kid's room is arranged with whole piece of open display cabinet to collect books and displays; different colors describe unique personalities of the residents.

公共空间串连客厅和餐厅的机能性，并以温润的木质色系铺陈，搭配浅色壁面及洞石材质，形塑出材质和色彩的丰富对比变化。主卧室利用简约沉静的色调沉淀环境氛围，窗边采用白色百叶窗，依据阅读或休憩需求，让屋主能随时弹性调整室内光度，回到卧室就能舒适放松。除了一般的衣物储藏空间，儿童房也安排整面开放展示柜体，可以收纳书籍和展示品，并由不同的色彩，摹画出属于居住者的独特个性。

KELLY HOPPEN DESIGN MASTERCLASS

铁粉最爱 质感居

The Favorite Textured Residence

设计公司：千彩胤空间设计有限公司
Design company: Cty Space

设 计 师：李千惠
Designer: Peggie Lee

项目地点：台湾高雄
Location: Kaohsiung, Taiwan

项目面积：182m²
Area:182m²

主要材料：进口薄板、核桃木钢刷木皮、铁件、烤漆玻璃 、实木地板等
Main materials: imported thin board, walnut steel brush wood veneer, iron, paint glass, solid wood floor, etc.

摄 影 师：刘欣业
Photographer: Xinye Liu

DESIGN CONCEPT
设计理念

The designer regards respecting the life of residents as the design principle of space planning, masters rights to use every space and designs of interior details, presents resident's personalities in the space and integrates them with creative designs.

In this project, the typical traditional conservative parents meet with their son who loves modern designs, which is not only concept conflict between the two generations, but also the thought collision between conservation and innovation. The owner who is a doctor is a big fan of the designer and expects a "surprising" modern home. So starting from the owner's personality, the designer abandons light color wood veneer, stone and brush paint and outlines a new era residence which amazes the elders through creative elements and designs.

以尊重居住者的生活，作为空间规划的设计准则，把握每一个空间的使用权及室内细部环节的设计，在空间展示您的个性，也融汇我们的创意设计。

本案是典型的传统保守父母，遇到喜爱现代有设计感的儿子，不仅是两代间的观念冲突，也是守旧与创新的思想冲撞。从事医生工作的屋主一直是设计师的铁粉，期待创造有“惊艳感”的现代住家。因此设计师以屋主个性为出发点，舍弃其不喜欢的浅色木皮、石材与刷漆，透过创新元素与笔触，勾勒出长辈也惊叹不已的新时代居宅。

◇ DECORATIVE MATERIALS
装饰材料

The spacious and clear public space in an open way is bathed in the warm sunshine fallen through wood shatters. The dark color wood veneer which connects the porch cabinet and extends inwards turns upward and cuts orderly with the beam lines to outline the dining room. TV wall in the living room adopts natural stone materials to present natural layering. The master bedroom continues the technique in public area and integrates bathroom into stone material wall. Netherlands scrape wood floor and wall modeling hide the master bathroom and reach the owner's expectation of rough taste as the wooden pallet.

敞朗清透的公共空间，以开放型态共沐在筛落木百叶帘映入的暖阳中，透过接续玄关柜体向内延伸的深色木皮元素，向上转折切齐梁线包覆出餐厅分界意象，客厅电视墙另采自然石物表现自然层次。主卧房延续公领域将卫浴融入石物壁面的手法，运用荷兰手刮木地板壁面造型隐藏主卫浴规划，同时达到屋主期待如同木栈板般的粗犷味道。

厨房
玄关
餐厅
客浴
客房
主浴
书房
客厅
客房
主卧

◇ SPACE PLANNING 空间规划

The area behind the sofa in the living room is planned as a party corner, echoing with the previously bought indigo blue sofa. Writeable paint glass door piece is set in front of the display cabinet. Simplified space lines achieve the owner's design thought of "the clearer the better" and widen the depth of field.

利用客厅沙发后空间规划一方聚会段落，并呼应事先购入的靛蓝沙发色系，在展示柜体前安排可书写的烤漆玻璃门片。简化后的空间线条，实现屋主“越干净越好的”的设计想法，进而拉阔场域景深。

◇ NATURAL LIGHTING 自然采光

Jumping out of the traditional space configuration thought, the end of the bed in the master bedroom uses black glass to build the half-penetrable dressing room. The position of door piece is the source of bright lights, which can create rich space visions. The tall floor area which originally plans to be the dressing room is used as a study. Concise desktop and high chairs near the window create a tailor-made sunny Starbucks reading area.

跳脱传统制式的空间配置想法，主卧床尾处采用黑玻建构半穿透感更衣室，门片的开阖或是光源的明亮，皆可变化丰富的空间视感。而原规划为更衣室的架高地板区则作为书房使用，临窗区的简洁桌面与高脚椅，更是量身打造的星巴克感阳光阅读区。

艺术自然 合韵相生

Art and Nature Perform Together with Charms

项目名称：国泰一品史公馆
Project name: Guotai Yipin Shi Residence

设计公司：京玺国际股份有限公司
Design company: EXCLAIM UNITED CORP.

设 计 师：周燕如
Designer: Joy

项目地点：台湾台北
Location: Taipei, Taiwan

项目面积：248m²
Area: 248m²

主要材料：进口磁砖、天然薄岩板、超隔音玻璃、老橡木实木地板、植生墙等
Main materials: imported tile, natural thin rock beam, super soundproof glass, old oak solid wood floor, green wall, etc.

DESIGN CONCEPT
设计理念

Guotai Yipin building is located in the green alleyway in Dunhua South Road, tranquil and leisure. It uses city environment to meet the needs of residents and accumulates perception experience of visual sense. The field correlates the environment and reflects the comfort needed by inner hearts by open space. Light and shadow change every day and night; sunshine and green plants flow continuously. The comfortable feeling is born from heart. Instantaneously, the space becomes the concrete embodiment of free thought and thus the spacious layout comes out.

Life interest is injected into the residence. The beam structure constitutes a three-dimensional effect. Under the modern geometric line array, it sets an opening layout and puts art gallery concept into the corridor. The display functions are distributed in the layout to encounter with residents and to endow the space with fun. Residents can enjoy afternoon sunshine near the bed. Living a lazy life, leisure and happiness are self-evident.

国泰一品大厦位于敦化南路盎然绿意的巷弄里，静谧悠闲，借由城市环境回应居住者的生活所需，累积视觉感官的感知经验。场域照应环境，以开阔空间映射心里所需的舒适感。光影日夜变换，阳光绿植，川流不息；惬意感受，由心而生。瞬时，空间成为自由意念的具体化身，阔朗格局因势而生。

生活兴趣，融于居所。梁体结构型塑三维效果，在现代几何线条的数组下，分割动线开口，廊道植入艺廊概念，展示机能遍布于动线里，与居者交集相遇，赋予遐想空间的趣味。临床空间以享受午后暖阳为乐趣，慵懒生活，闲适美好不言而喻。

◇ INTERIOR VIRESCENCE 室内绿化

The public and private outdoor balcony uses green wall design and flow of light and shadow to present exuberant vitality. The visual movement of sitting or lying sets humanized height, combines modern space tone with the thickness of material texture, feels the rich passing of nature and time and brings different viewing experiences from green plants to life.

环伺公私领域的室外阳台，透过植生墙的设计，通过光线的流转引入室内，挹注盎然生气。经由或坐、或卧的视觉动线，设定人性化的高度，以现代空间基调结合材质肌理的厚度，感受丰富的自然与时序的推移，带来室内绿植给予生活的不同观赏经验。

◇ NATURAL LIGHTING　自然采光

The arrangement of intelligent system coordinates with large area of glass French window and spacious field. Light and shadow naturally adjust according to time, place and people to enter into interior or stay outside. The long, short, heavy and light changes present rich and layering structure, derivative rhythmic space sense and abundant validity and make the residence elegantly become a castle beyond architecture, art and nature.

智能系统的安排，配合大面积玻璃落地窗和朗阔的场域空间，光影因时、因地、因人所需，作自然调节，导入屋内或停留在外。在其长短浓淡的变化，体现层层丰富的结构，衍生充满律动的空间感、丰沛的生命力，让居宅优雅的成为跨越建筑、艺术、自然的堡垒。

◇ LIGHTING DESIGN 照明设计

Opening the curtain of night, there is less illumination of outside natural light and shadow but more warm and vernal humanistic light and shadow to light the happy temperature of family. The public areas of living room and dining room are clean and clear; the hidden strip lights or distracted orderly tube lights add night atmosphere and meet the need of illumination. The droplight of the dining room condenses the dining atmosphere. The private area of bedroom adds a bedside night light to illuminate the space and to add warmth for the heart.

拉开夜的帷幕，少了屋外自然光影的照射，却有温暖和煦的人文光影点亮家的幸福温度。客厅餐厅公共领域洁净爽朗，以暗藏的灯带或者分散有序排列的筒灯为场域增添夜晚的气氛及满足照明需求，餐厅吊灯凝聚就餐氛围，私领域卧室则增加床头夜灯，点亮一方，增添心灵的暖。

泰然自若 自在璞真

Calm and Composed, Comfortable and Plain

项目名称：泰然 璞真
Project name: Calmness and Plain

设计公司：创研空间
Design company: Create+Think Design Studio

设 计 师：何俊宏
Designer: Arthur Ho

项目地点：台湾台北
Location: Taipei, Taiwan

项目面积：284m^2
Area: 284m^2

主要材料：石皮、天然板岩、胡桃木皮、香杉、壁纸、胡桃木地板等
Main materials: stone skin, natural slate, walnut wood veneer, cedar, wallpaper, walnut wood floor, etc.

摄 影 师：李国民
Photographer: Kuomin Lee

DESIGN CONCEPT
设计理念

Inheriting the design concept of "thought of life and touch of heart" adhered by Create+Think Design Studio, the designer sets "plain" and "comfort" as the design concepts, uses materials such as natural slate, stone skin and walnut wood, expresses and sublimates the charm and tone of the space through the intersection of materials and combinations of lines, and creates a spiritual enjoyment in the natural and indifferent space.

The designer combines Chinese elements with modern minimalism. The whole project presents elegance and Zen of neo-Chinese style. "The Zen believes, the Zen Buddhism is in the world, holding a common heart and paying attention to the present; so I hope that residents can refine 'elegance' from everyday living details; 'elegance is common' is the mood I want to convey." Says the designer.

秉承创研空间一直以来"有生活的思考、有心灵的感触"这一设计原则，设计师以"璞真"、"自在"为设计理念，通过天然的板岩、石皮、胡桃木等材料的配置使用，在材质的穿插和线条的组合中，怡然自得地表达和升华了空间的气韵格致，在自然淡泊的空间中成就一种充盈于心灵的享受。

设计师将中式元素与现代简约主义相结合，整体呈现出新中式的风雅和禅意。"禅宗认为：佛法在人间，持平常心，注重当下，所以希望居住者们能从日常的生活细节中提炼出'风雅'，'风雅处处是平常'就是我想传达的心境。"设计师如是说。

CLOSE UP

◇ DECORATIVE MATERIALS 装饰材料

The designer uses natural materials such as stone skin, slate, solid wood, wood veneer, complemented by Chinese window lattice. Sun light passing through the void net curtains activate the space, creating a calm and composed, carefree and comfortable plain living environment.

设计师运用石皮、板岩、实木块、木皮等天然材料，佐以中式窗棂，网状透空窗帘过滤阳光的加入则活跃温暖了空间，创造令人可泰然自若、悠然自在生活的朴真环境。

CLOSE UP
KELLY HOPPEN

主卧
卧室
书房
佛堂
客厅
餐厅
厨房

◇ SPACE PLANNING 空间规划

The space is divided into two large areas, the public area and private area which are connected by the front hall. The public area is composed of two units, the living room and dining room which are connected into a large space by the wood ceiling shaped like a rectangular lampshade. The private area includes the master bedroom, study and living room, quiet, comfortable and cozy.

空间分成公领域跟私领域两大区域，以前厅相连结。公领域由客厅、餐厅两个单元组成，以一似长方型灯罩的木制天花造型串连成一大空间；私领域则包括主卧室、书房、起居室等空间，静谧、自在、舒适，兼而有之。

◇ NATURAL COLORS　自然色彩

In the public area, two sides of the double-sided cabinet in the hub position are walnut wallboard with dark and light grains, extending to the right and becoming the basic tone of this area. The side near the French window at the end of the sitting room is the gray stone skin wall embedded with cedar wood. French windows on two sides of the wall match with void net curtains, forming the interesting light and shadow, which is the side view of the large space.

公领域中，位于枢纽位置的双面柜两侧为深浅木纹相间的胡桃木壁板，向左右延伸，成为此区的基底色调。末端客厅邻落地窗侧是灰色石皮墙面嵌着香柏木块，墙面两侧落地窗是网状透空窗帘，形成趣味光影，是此大空间的端景。

wolterinck's world OUTIIN

生命的光 Light of Life

设计公司：禾筑设计
Design company: Herzu Design

设 计 师：谭淑静
Designer: Tam

项目地点：台湾台北
Location: Taipei, Taiwan

项目面积：84m²
Area: 84m²

主要材料：进口薄片砖、不锈钢材与铁件、涂装木皮板、大理石、进口磁砖、丝绷布、进口壁纸、风琴帘等
Main materials: imported thin tile, stainless steel, iron, coating wood veneer, marble, imported tile, silk fabric, imported wallpaper, organ shade, etc.

摄 影 师：吴启民
Photographer: Qimin Wu

DESIGN CONCEPT 设计理念

Designs of this project break the limitations of style framework, implant sanative space concept, return to people's psychological perception, contemplate the relationship between human and space and explore residents' real life needs. The owner wants the house to be vacant and hopes that designs can enlarge the room; "vacant is not empty without anything" but to use perception and vision of the space by modern technique to design. "Less is more", the designer gets rid of common normal space space definition technique, removes unnecessary decorative vocabularies and endows the field the largest scale and elasticity.

The design concept "light of life" is used throughout the whole house. Light not only is used to create visual effects, but also influences the layout and layering of the space. The metal ceiling which connects the public area with private area combines with hand-sprayed acrylic light groove. At two ends of the layout, biblical passages are translated into password type, restrainedly depicting the inside meaning.

超脱风格框架的局限，植入疗愈空间概念，回归人的心理感知，审慎思考人与空间的关系，探讨居住者真正的生活需求。屋主要求屋子要空，希望通过设计把屋子放大，但“空不是无物”，而是将空间的感知与视觉以现代主义的手法来设计。“Less is more”舍去以往常见制式的空间定义手法，去掉无谓的装饰性语汇，赋予场域最大的尺度与弹性。

“生命的光”的理念贯穿全室，光不仅是用于视觉效果，在空间的动线与层次上也发挥效果，串连公领域与私领域的金属天花板结合手喷压克力灯沟照明，在动线的两端，以圣经章节转译成密码型式，内敛地刻画其中的含意。

◇ DECORATIVE MATERIALS AND SPACE PLANNING 装饰材料、空间规划

The living room and dining room form a good annular comfortable layout, jumping out the old furniture placement pattern. Instead, from the human point of view and thinking different situations of everyday life, the furniture is put in a humanistic configuration with temperature. The basic tone of the space as the base has extension effects. Different materials have similar colors, for example, faint glossy thin tiles or light gray paint surfaces with temperature which stock from walls and cabinets, foil the upper modern glossy stainless steel and fluoride acid mirror, with light groove outlined in the plain base, condensing the layout, materials and lights.

The designer returns to the user's perspective, delicately thinks every detail when cooking such as movement, storage location, using function and lighting requirements; material applications continue tonality of the whole space tonal. The back balcony door pieces use large piece of shutters for visual integration, clean and neat, and use lights from the window gap to create a natural atmosphere.

客餐厅保持了良好的环状舒适动线，跳脱陈旧的家具摆放形式，改而从人为出发角度，思考着不同生活使用情境，摆置出具有人的温度的配置。空间的基础色调恰如其分的作为基底与延展的效果，相似色感却皆不同材质，如隐隐光泽的薄片磁砖或那具有手感温度的浅灰色漆面从墙面、柜门等部分堆栈，衬托其上具有现代光泽感的不锈钢与氟酸明镜材质，并在素底上勾勒出灯沟，使动线、材质、光线得以凝聚。

回归到使用者的角度做思考，细腻思考烹调时的各项需求细节如动线、收纳位置、使用机能、灯光需求，材质运用延续整体空间调性。后阳台门片采用大面百叶窗作视觉整合，干净利落，并利用光线从窗隙中透出，营造自然氛围。

◇ LIGHTING DESIGN 照明设计

The corridor in the public area of this project stresses light design. Oblique lines and shining lines which extend to the elevation and represent lights use customized places to illuminate the triangular light box, in addition to strips average lighting. Entering from the gate, the left white wall has some light grooves to guide and create atmosphere. The designer quotes a sentence from "John chapter 8, section 12" of the Bible "I am the light of the world. Whoever follows me will never walk in darkness, but will have the light of life." At two ends of the corridor, this sentence uses Morse code to transform the telegraph language into elevation lighting design.

此案公共区走廊以光作为设计重点，斜线条及线条延伸至立面代表光的照耀线条，以客制的点照明三角形灯盒，加上条状的平均照明结合。从大门进来左边留白墙面开了一些光沟，用光做引导及营造氛围的目的。我们引用《圣经》里的"约翰 8 章 12 节"中的一句话"我是世界的光，跟从我的，就不在黑暗里走，必要得着生命的光！"在走廊的两个末端，把这句话用摩斯密码的方法，把电报语言转化成立面照明设计。

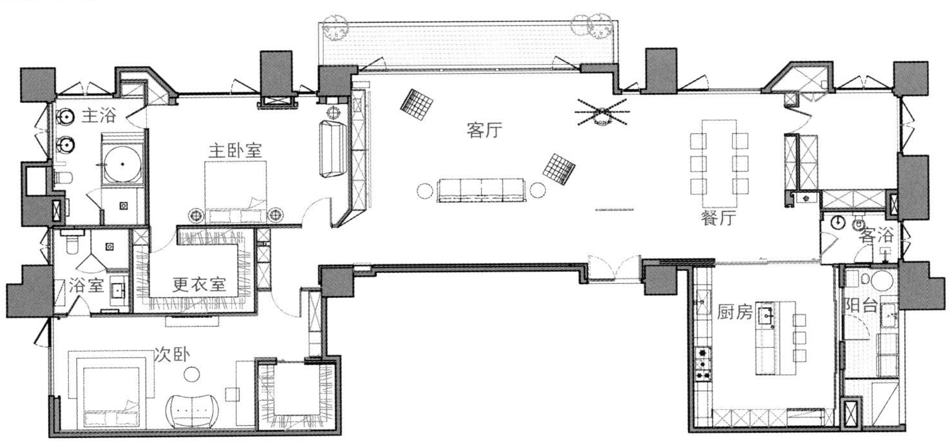
主浴
主卧室
客厅
餐厅
客浴
浴室
更衣室
厨房
阳台
次卧

◇ NATURAL COLORS 自然色彩

Deep to the bedroom, the designer explores people's perceptions to materials and uses Thai silk fabric to cover the wall; the meticulous tactility reflects soft light gray luster, presenting a quiet space feeling rather than a presumptuous exaggeration. The stainless steel wall lamps embedded in the wall function as outline decorations and amuse life situations. In the evening, the light provide convenience for people who walk in the space.

In the kid's room, the designer uses composition technique to plan the proportion of the space and chooses blue wall board, white wood veneer and lamplight layering to present a lively space feeling. The bathroom extends the lively and clear tone of the bedroom; geometrical arranged totem of wall bricks make the wall feel like a whole piece of canvas which endows the space with lively tone.

回归到卧房深处，探讨人对材质的感受，采用泰国丝绷布做墙面包覆，细致的触感反射出柔和的浅灰色光泽，呈现静谧的空间感受，而非喧宾夺主般的夸大喧闹。壁面上嵌入的不锈钢体壁灯，同时具有轮廓修饰与生活情境功能，可在夜晚中，微微点亮动线，帮助走动便利。

小孩房采用构图手法规画空间中的比例，利用蓝色壁板、白色木皮材质与灯光层次，呈现轻快的空间感受。浴室延伸卧房轻明调性，壁砖构图式的排列几何图腾，墙面就像整张画布般，为空间活泼上彩。

凝聚·家 Cohesion and Home

设计公司：共禾筑研设计有限公司
Design company: Nison Space Design Co.

设 计 师：陈煜棠
Designer: Yu-Tang Chen

项目地点：台湾台中
Location: Taichung, Taiwan

项目面积：350m²
Area: 350m²

主要材料：木皮、石材、玻璃、岩板、皮革布、进口家饰布等
Main materials: wood veneer, marble, glass, rock beam, leather panel, imported cloth, etc.

摄 影 师：钟崴至摄影工作室
Photographer: Weimax Studio

DESIGN CONCEPT
设计理念

The designer uses innovative thoughts and concise technique, conforms to the diversity of space and arranges better circulation to create layering of the space. He also uses simple materials, divides rigorous proportion and applies the uniqueness of the space itself to interpret the excellent dialogue between the space and the user.

This project is a transparent villa in modern style. Based on pure white, it applies diverse materials and rich line changes to inject exquisite connotative texture into the house. The designer emphasizes the essence of life rather than the superficial form of space to make it and its user's life emotion resonate and to highlight the meaning of the existence of the space. In the design process, the designer not only defines the space, but also creates the meaning of it.

以创新的思维与简洁的手法，因应空间多样性，配置良好的动线规划，创造空间的层次感；以简单的材料运用，严谨的比例分割，运用空间本身的独特性，演绎出空间与使用者的精彩对话。

本案以现代风格为主题的透天别墅，纯净的白色为空间的基底，搭配多元的材质与丰富的线条变化，为家注入细腻的内涵质感。重视生活的本质，而非空间的表面形式，让空间与使用者生活情感产生共鸣，突出空间存在的意义。在设计的过程中，不仅仅只在定义空间，也在创造空间的意义。

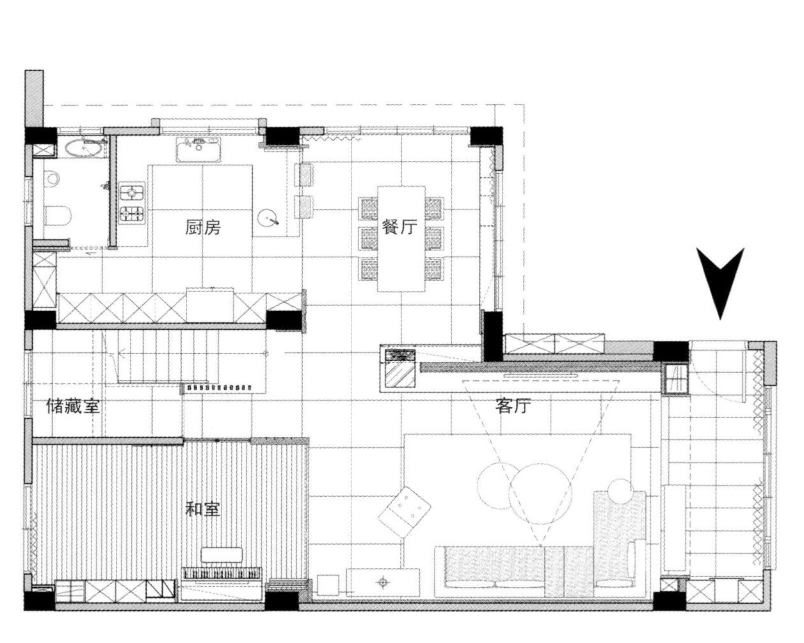

厨房
餐厅
储藏室
客厅
和室

◇ DECORATIVE MATERIALS AND SPACE PLANNING 装饰材料、空间规划

The sofa background wall uses world map as the main axis of space design, creating a space atmosphere with high international vision and becoming the visual focus of the space. In the living room, gray marbles of the TV wall use line segmentation in accurate proportion to add variability to the wall. The left side of TV wall connects with the embedded fireplace and open wall cabinet, showing magnificent momentum of the mansion.

The iron and laminated shelf above the fireplace extends to the dining room, providing abundant display space. On the stair side, iron grilling matches with thin rock beam with beautiful cutting to show concise sense of design. The multi-functional room with high floor uses iron folding door as the partition, which can be used as a piano room and a guest room when there is a visitor.

沙发背墙以世界地图作为空间设计的主轴，营造出具有高度国际视野的空间氛围，也成为空间中的视觉焦点。客厅电视墙的灰色系大理石，以精准掌握的比例线条分割，增添墙面的变化性，电视墙左侧联结至嵌入式壁炉与开放式吊柜，展现豪宅恢弘的气势。

壁炉上方以铁作及层板置物架延伸至餐厅，提供丰富的展示空间。楼梯侧以铁作格栅呈现，搭配漂亮分割的薄岩板，展现简约的设计感。地板垫高的多功能室以铁作折门作为区隔，除当琴房使用，在客人来访时可充当客房。

◇ NATURAL COLORS 自然色彩

The black concave broken lines of the ceiling are another highlight of the space design and start from the porch to the living room, dining room and kitchen, outlining amorous feelings of the ceiling and wall. Smooth lines also add layering to the field. Black lines extend to private areas upstairs. The ceiling in the reading area with abundant books, black iron hangers in the master bedroom and black iron pieces in the entrance of the boy's room echo with elements in the public area and create a living atmosphere in modern style.

天花的黑色凹折线条是空间设计的另一亮点，从玄关处即开始蔓延，延伸至客厅、餐厅及厨房，勾勒出天花板及壁面的万种风情，流畅线条也增添场域的设计层次。黑色线条的设计元素延伸至楼上私领域，丰富藏书阅读区的天花、主卧更衣的黑铁吊架、男孩房入口的黑铁件，不仅与公领域的元素呼应，也创造出现代风格的居家氛围。

主卧室
更衣室
阳台
书房
阅读区

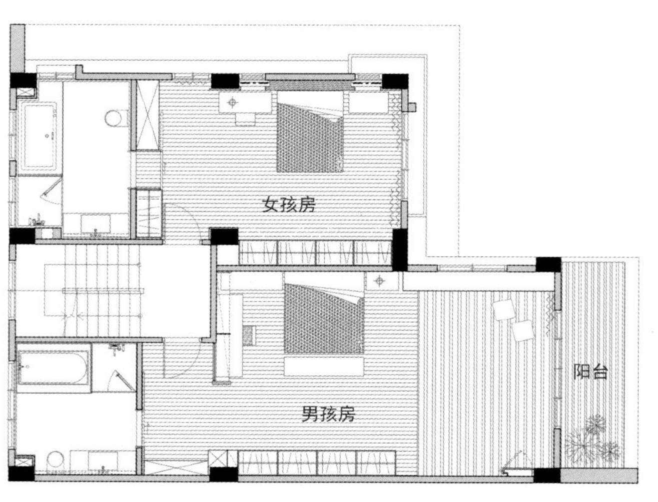

女孩房
阳台
男孩房

山林晓居

Mountain Forest Residence

设计公司：润泽明亮设计事务所
Design company: LIANG-DESIGN

设 计 师：林宥良
Designer: LIANG

项目地点：台湾新北
Location: New Taipei, Taiwan

项目面积：84m²
Area: 84m²

主要材料：超耐磨木地板、实木贴皮、特殊壁砖、角马赛克、地铁砖、台湾海棠花玻璃等
Main materials: super durable wood floor, solid wood lagging, special wall brick, angle Mosaic, subway tile, Taiwan begonia glass, etc.

摄 影 师：MD PURSUIT
Photographer: MD PURSUIT

DESIGN CONCEPT
设计理念

The old house is charming, precisely because it contains memories of the past. Overhauling the old space is changing the pattern to take its advantages. But for the most important residents, partially reshaping appearance of the old house, retaining traces of old memories, presenting their habits and preferences in the environment through modification can create the truest appearance of home. The designer finds out advantageous mountain scenery of this 40-year-old house, integrates wood texture and art collections which the couple loves, brings window scenery into interior, uses wood building materials to connect interior with exterior and creates a leisure and comfortable living environment.

Living here, you can feel the energy and pulse of the mountain forest. In the rainy day, the mountain is obscure; in the sunny day, the sun falls inclined. Breath in or out, you can be carefree and comfortable.

老屋之所以迷人，正是因为它蕴藏了过去的记忆，翻修活化旧有的空间，改变格局发挥住宅的优势，而对于最重要的居住者来说，部分重塑老宅面貌保留过去记忆的脉络，将居住者生活习惯、喜好透过改造后呈现在环境中，才是最真实的家的样貌。设计师发掘这间 40 年老屋得天独厚的山景，糅合两夫妻喜爱木头质感与艺术收藏，将窗景引入室内，木系建材连接屋里屋外，构筑悠游自在的居住环境。

居住其中，就能感受到山林的活力与脉动，雨天山岚迷幻，晴日斜阳轻挪，呼吸吐纳，悠哉游哉。

DALI
Meier
DRAWING PEOPLE
Kerry Hill

◇ DECORATIVE MATERIALS 装饰材料

Although multiple building materials are used, whether it is traditional or modern, Oriental or Western, all uses “warm” and “primitive” tonality to shape this residence. The ground is covered with Germany MASTER North America light color super durable oak floors. The kitchen walls use begonia frosted glasses transformed from Spanish tiles and system cabinet combining with old articles. The bathroom uses black and white subway tiles with glass texture and hexagonal Mosaic tiles. Air brick shows the taste of the owner, and creates a harmonious atmosphere.

虽然使用多元建材，但不论是传统或是现代、东方还是西方，都紧扣着“温润”、“原始”的调性去形塑此住宅。地板是德国 MASTER 北美浅色橡木超耐磨地板，厨房壁面为西班牙花砖与系统柜结合老件改造的海棠花毛玻璃，浴室使用玻璃质感的黑白地铁砖与马赛克六角砖；空心砖是屋主的品味，营造和谐的氛围。

餐厅
厨房
书房
客厅
卧室
浴室
玄关

◇ NATURAL COLORS 自然色彩

The whole space is based on green and wood color and uses the most natural colors. The green of window scenery is introduced into interior, collocating with green bonsai, sofa, lake green paint in the bathroom and so on. In order to avoid too many colors in the space, the designer gives priority to natural color and texture of logs and naturally paves them in every corner of the house.

全室采用绿色和木色为主，利用最自然的色彩——窗景的绿引入室内，搭配绿色盆栽、沙发、浴室湖水绿涂料……避免过多色彩混淆空间，将主角让给原木的自然色泽与纹理，自然铺陈居家每个角落。

◇ SPACE PLANNING 空间规划

The designer opens the kitchen to connect with the living room, introduces more lights and air and keeps the balcony space. In addition, the designer also appropriately plans the indoor and outdoor space proportion, readjusts the former balcony and piles creamy white stone in the cement fence. You can see veins of the old house outdoor. Since there is only one balcony, the designer integrates viewing and working functions, introduces pumping appliance of the washing machine under the cabinet to the balcony and maximizes space functions.

设计师将厨房打开，与客厅连接，引入更多的光与空气，保留阳台空间。除此之外，设计师也妥善规划室内外空间比例，重新调整阳台，将水泥围栏砌上米白色皿石子，在户外也能重现老屋的脉络。由于只有单一阳台的配置，将观景与工作阳台的机能整合，将橱柜下的洗衣机排水设备引到阳台，发挥空间的最大功能。

微奢风
LIGHT LUXURIOUS STYLE

Design Concept
设计理念
Decorative Materials
装饰材料
Natural Colors
自然色彩
Natural Lighting
自然采光
Interior Virescence
室内绿化
Bringing Scenery Into House
引景入室
Space Planning
空间规划
Lighting Design
照明设计

简·致 Simplicity and Perfection

项目名称：未来
Project name: Future

设计公司：近境制作
Design company: Design Apartment

设 计 师：唐忠汉
Designer: TT

项目地点：台湾台北
Location: Taipei, Taiwan

项目面积：228m²
Area: 228m²

主要材料：石材、不锈钢金属、镀钛金属板、橡木染灰等
Main materials: stone, stainless steel, titanium plate, oak dyed gray, etc.

设 计 师：李国民
Designer: Kuomin Lee

DESIGN CONCEPT
设计理念

Structure misplacement presents the concept of interior architecture while structure imbedding differentiate main elements in the space.

The space uses imbedding of layered and overlapped structure to naturally divide the main and deputy field properties so as to present the design concept that interior design is integrated with architectural elements. In this project, layout planning between each structure is very meticulous. The designer creates a natural and comfortable living space where different areas are coherent. With dynamic flow and open vision, this space is injected with some future technologic senses, which has unique and exquisite designs and is full of cultural appeal.

以量体错置展现室内建筑的概念，以量体置入区分空间的主要元素。

空间以量体层覆交迭的置入，自然地分划主副场域性质，呈现出室内设计融入建筑元素的设计概念。在这套案例中，每一个量体与量体之间的动线规划皆十分缜密，设计师打造了一个各区域相贯通的自然、舒适的生活居住空间。场域流动，视觉开阔，而且在这样一个空间中加入些许穿越时空的未来科技感，既设计独特精巧，又富有文化感染力。

◇ DECORATIVE MATERIALS AND SPACE PLANNING 装饰材料、空间规划

In the functional division of interior space, the designer uses stainless steel plate in the ceiling to strengthen the relationship between subject and object boundary, connects the living room with dining room and adopts Western bar design to strengthen the spacious and sedate visual sense of main spaces. The public area is static and dynamic. Natural logs quietly stack up; the stone surface and wood color set off each other by contrast. In the bar area, the chiseling of wall material is as vertical and straight as the waterfall, contrasting directly with the calmness of the stainless steel ceiling. The gallery connected with kid's room is based on white and uses progressive modeling to imply the kid's innocence and best wishes for growth in the future.

在室内空间的功能划分中，设计师以天花采用不锈钢板强化主客关系分界，并将客餐厅领域连结，融入西式吧台设计，由此更能加强主空间宽敞、沉稳的视觉感。公共区域静中有动，极具自然感的原木安静地摞列放置，石台面与木之色相互陪衬。吧台区壁材的凿切如瀑幕般中垂直下，与不锈钢天花的平静产生直接性对比，连接小孩房的长廊以白为肌底，并以层迭渐进造型隐喻孩童的无瑕和对成长未来的祝福。

◇ NATURAL COLORS　自然色彩

Entering into the master bedroom, the multi-functional long table uses color and luster of landscape to obscure the main and complementary field properties which echo with each other through different materials and structures. At the same time, the designer abandons complicated decorations, uses extremely simple lines to outline and creates a private space which is neat and tasty in sense and more elegant in thought. The designer uses different grains such as the warmth of wood, the magnificence of stone and the coolness of metal to illustrate the owner's tenderness at home, expectation of career and firmness of a person, creating the dialogue between residents and space through the transformation of three temperatures.

步入主卧，多功能的长桌以山水色泽模糊主副空间性质，透过不同的材质与量体间达到相互呼应。同时摒弃复杂的装饰，用极简的线条勾勒，营造感官上整洁、品味和思想上更为优雅的私密空间。木质的温润、石材的壮阔以及金属的冷冽，运用不同纹理所阐明的心境体现业主居家之温情、事业之展望以及为人之果决，透过三种温度的转化型塑出居者与空间的对话。

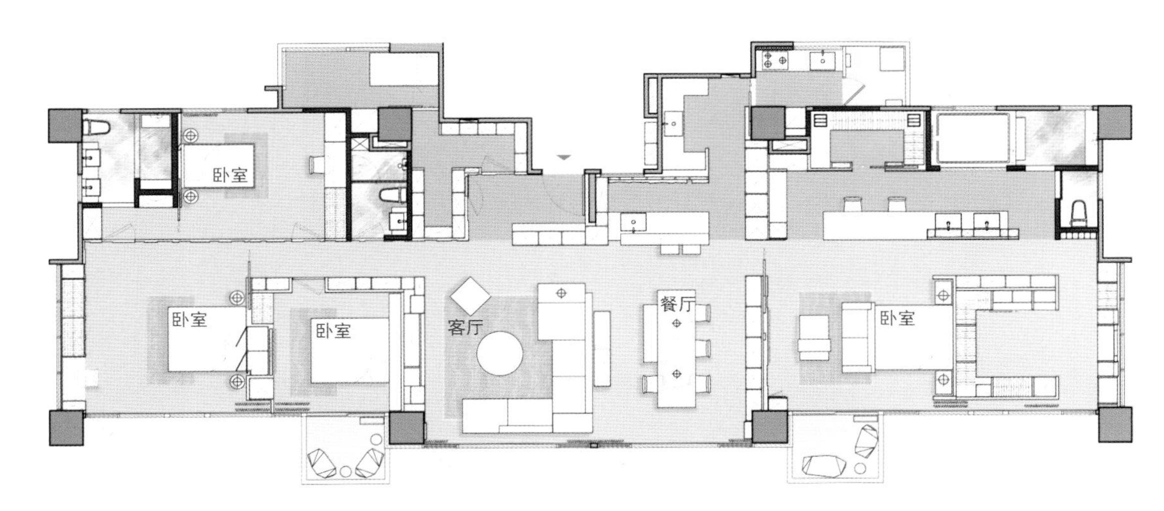

设计公司：仝育空间设计有限公司
Design company: TONGYU Space Design CO., Ltd.

设 计 师：庄媛婷、郑瑞文
Designers: Chuang Yuan-Ting, Cheng Jui-Wen

项目地点：台湾台南
Location: Tainan, Taiwan

项目面积：550m²
Area: 550m²

主要材料：铁件、灰玻、黑镜、烤漆、裱布等
Main materials: iron, gray glass, black mirror, stoving varnish, fabric, etc.

摄 影 师：利德凯国际空间摄影团队
Photographer: LDK Photography Studio

深喻恬静 自然共生

Profound and Tranquil, Natural and Symbiotic

DESIGN CONCEPT 设计理念

Located in the cluster of single-family residences in Tainan Humei Distrait, this project is adjacent to the garden with vast lush green trees and ecological pool and has perfect living conditions. The design of the gray and white tone makes the residence quietly hide in the wide surrounding scenery in a low-key way. The exterior is dotted by greens in the terrace, the balcony and the inner courtyard. The refined dry landscape Zen creates a tranquil corner to precipitate body and mind so as to create the slight natural atmosphere. The dialogue between them creates a symbiotic and syncretic relationship among people, life, environment and ecology.

Continuing the concept of gray white modern style, the interior uses concise and neat features to make full use of modern and natural elements in the entire field through the design technique of matching white metope with wood texture.

本案位处独栋式住宅群聚的台南湖美地区，邻近由大片蓊郁绿树与生态池打造的公园，拥有绝佳生活机能。灰白色调的设计，让居所低调地隐身于周遭广阔地景致中。外部透过露台、阳台及内院中庭的绿意点缀，以雅致的枯山水禅意，打造能沉淀身心的宁静一隅，借由创造以空间微塑自然的氛围，并通过两者间的对话，营造人、生活、环境、生态，彼此之间共生的融合关系。

延续外部以灰白色系现代风格为主调的理念，室内部分由掌握简约、利落的特质，透过留白墙面混搭木纹质感的设计手法，将现代与自然元素充分运用于整体空间场域中。

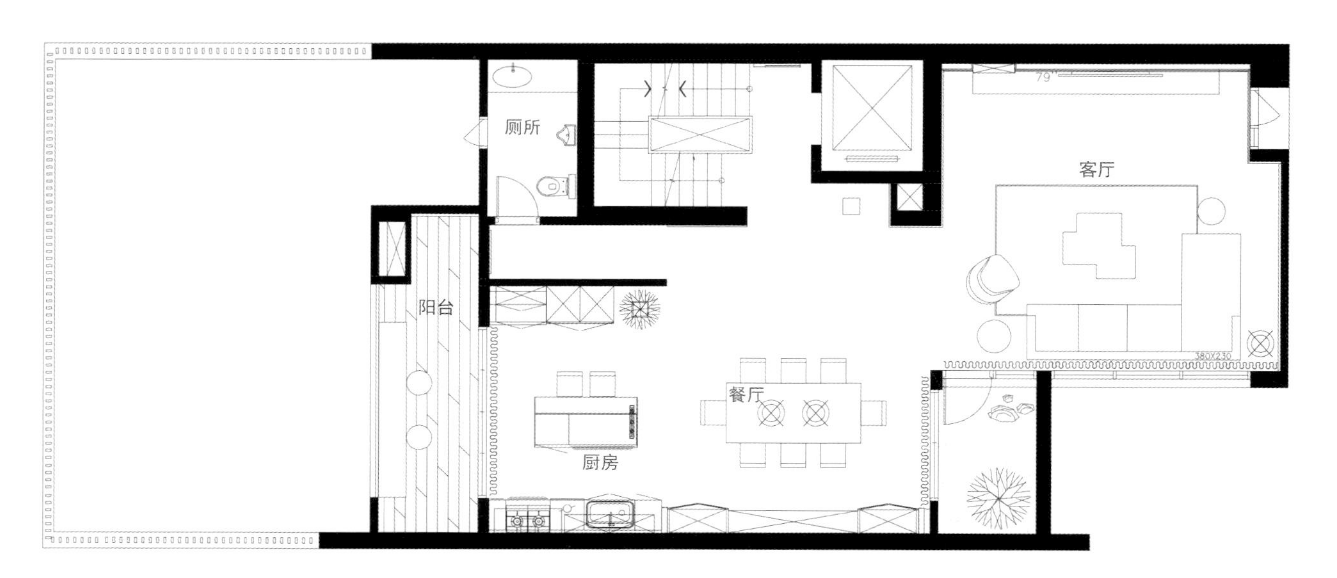
厕所
客厅
阳台
餐厅
厨房

◇ SPACE PLANNING 空间规划

In the first two floors which pursue public life functions mainly, the designers use modern art works with extremely Zen-like artistic conception to combine Oriental cultural connotations with Western contemporary style so as to create an Eastern and Western space atmosphere which integrates ancient and modern elements. In the upper three floors which pursue private life functions, the appropriate collocation of light and color creates a tranquil and moderate life atmosphere.

在以公共生活机能为主要诉求的一、二层室内，设计师以极具禅风意境的现代艺术品为核心，将东方人文底蕴与西方当代风格相组构，荟萃出东西合璧、纳古鉴今的空间氛围。以私人生活机能为诉求的三、四、五层室内，则以恰好的光线色调搭配，启发出安静温和的生活场域氛围。

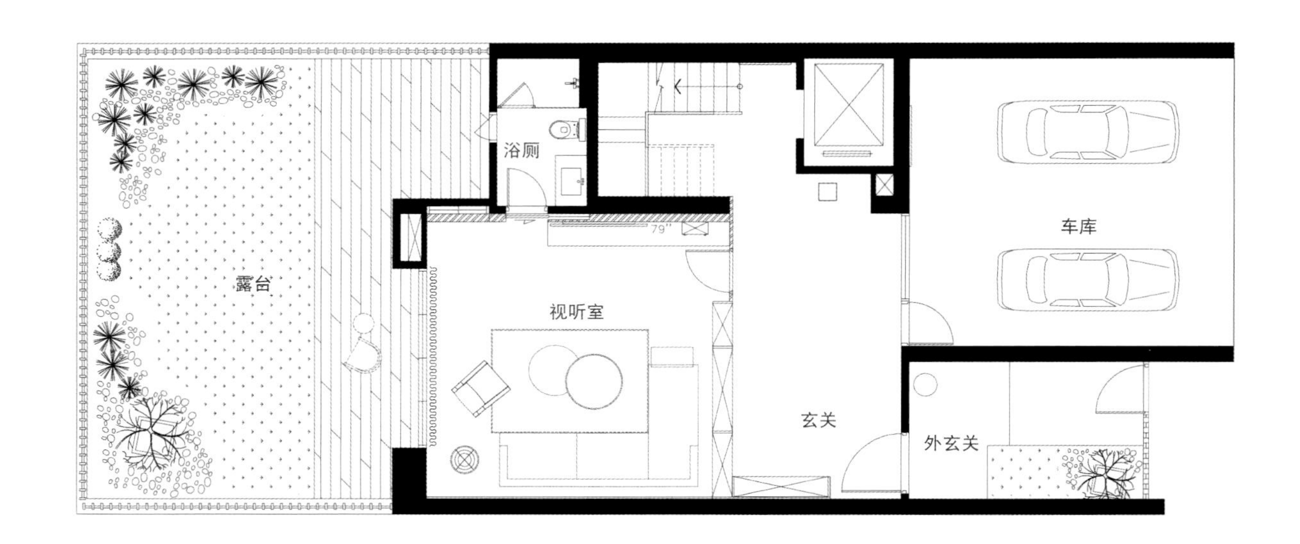

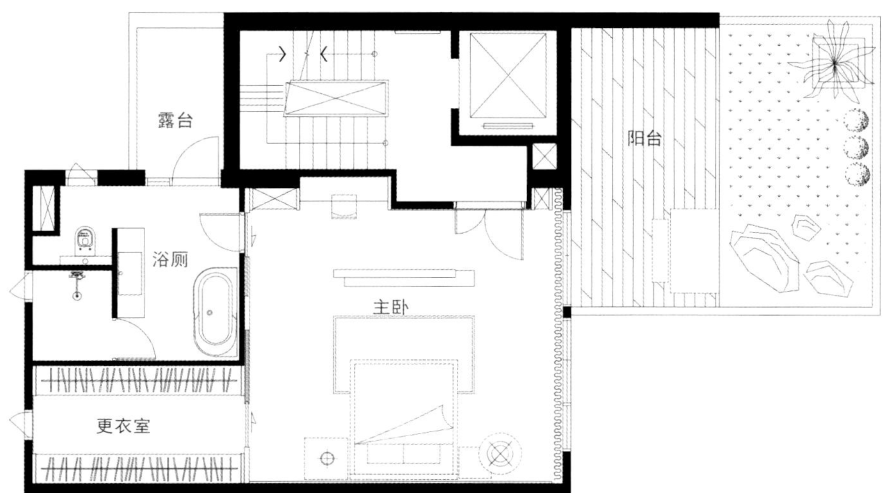
露台
阳台
浴厕
主卧
更衣室

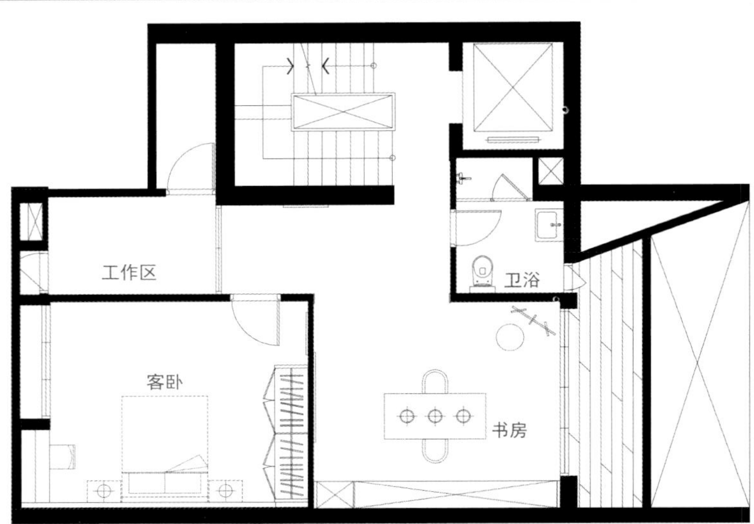

◇ DECORATIVE MATERIALS 装饰材料

In the living room, the stone main wall with magnificent grain and smooth texture which functions as the mountain and large piece of modern splash-ink painting as waterfall collocate with stone texture carpet and brown leather sofa, forming a vast and profound artistic conception as if in the natural forest. The dining and kitchen area behind the living room uses modern sculpture as the visual focus of the space, with plain designs and the applications of mirror and glass, bringing different interior and exterior scenery into it and creating a humanistic dining and socializing situation.

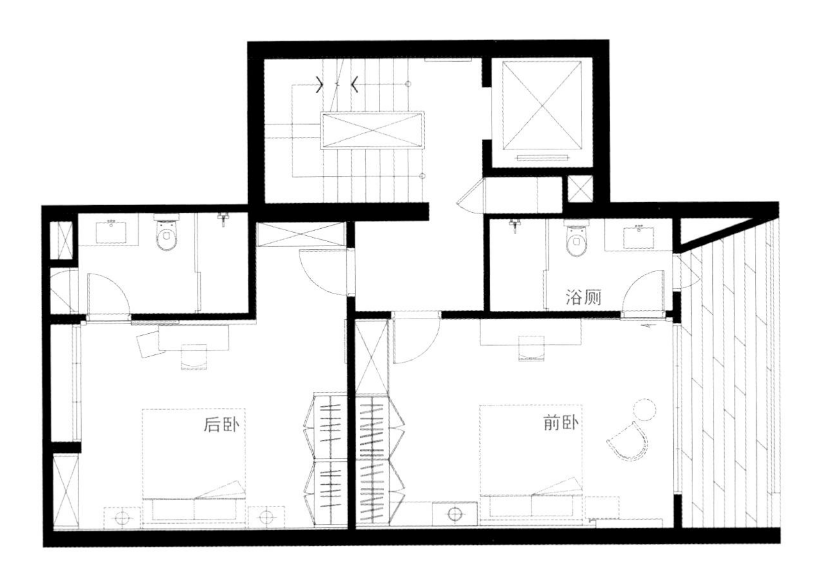

客厅内，由纹路瑰丽，质感细腻的石材主墙为山势，大面的现代泼墨挂画为水瀑，搭配同以石材纹理呈现的地毯及棕色调皮质沙发，形构出仿若置身自然山林般的辽阔、悠远意境。客厅后侧的餐厨区域，同样也以当代雕塑作为空间的视觉核心，搭配质朴的设计与镜面、玻璃等材质之运用，将不同的室内外景致援引其中、营造出充满人文风情的用餐、交谊情境。

◇ LIGHTING DESIGN　照明设计

Private bedrooms continue the concept of linking the space designs by using nature. In the master bedroom, the collocation of cold imitated stone main wall and warm wood grain ceiling creates a cold and hot, yin and yang conflict beauty. Other bedrooms are based on wood grains, with hanging lamp decorations as the unified visual focus; the warm space atmosphere creates and strengthens the sleeping and relaxing household features.

私密的睡房空间中，沿续以自然连结空间设计的理念，主卧部分透过冷调的仿石材壁面主墙与温润木纹天花板的搭配，辅以垂悬的不规则灯饰作为空间的视觉主体，形塑冷与热，阴与阳调和的冲突美感。其他卧房亦善用木质纹理的铺垫，以垂悬灯饰作为统一风格语汇的视觉核心，借由温润的空间氛围营造和强化睡眠、休憩的居家特质。

几何的淬炼
The Geometrical Refinement

设计公司：咏义设计股份有限公司
Design company: VERY SPACE INTERNATIONAL

设 计 师：刘荣禄、谢宜臻
Designers: Louis Liou, Yizhen Xie

项目地点：台湾台北
Location: Taipei, Taiwan

项目面积：310m^2
Area:310m^2

主要材料：锈铜砖、大理石、茶镜、墨镜、明镜、橡木染咖啡、橡木染灰、橡木染白、橡木喷白、柚木深刻纹、镀钛钢板、黑铁烤漆、乳胶漆、进口壁纸等
Main materials: rust copper tile, marble, tawny glasses, dark glasses, bright mirror, coffee dyed oak, gray dyed oak, white dyed oak, white sprayed oak, teak carved grain, titanium-coated steel plate, black iron stoving varnish, latex paint, imported wallpaper, etc.

摄 影 师：郭家和
Photographer: Jiahe Guo

DESIGN CONCEPT
设计理念

Different from general residential spaces, in this project, the designers use different elements to create different visual replacements and shape and present the theater space through the technique of geometry refinement so that residents can experience the living feelings with deep thinking atmosphere.

The designers challenge aesthetics again and again, try all the impossibilities and add asymmetry into many spaces in design to make space change along with thoughts. The designers are good at using charms of geometry to present extraordinary logical integration ability in the make-up process. Rich design elements collide and integrate in the field, creating a deep theater living space.

有别于一般住宅空间，设计师在本案中，利用不同元素造成不同的视觉置换，并透过几何淬炼的手法形塑、表现出这个剧场式空间，借以体验可沉浸在深度思考氛围的居住感受。

设计师一次次地挑战美学，尝试所有的不可能，在设计上加入许多空间的不对称性，让空间可以随着思维去改变。并善用几何的魅力，在拼凑的过程中，展现不凡的逻辑整合能力。丰富的设计元素互相碰撞、融合于场域中，打造出深具剧场性的居住空间。

THE HOTEL BOOK
MICHAEL JACKSON

◇ DECORATIVE MATERIALS 装饰材料

To present the depth of the space, different materials are used as vague partitions between fields. From the porch to the living room, special wood screen design replaces closed plate; the cylindrical surfaces use various materials to outline rich details. Different angles and different lights make this field present a unique story.

为展现空间的深度，各场域之间皆以不同材质作为隐约区隔的手法。自玄关至客厅，以木头材质的特殊屏风设计取代封闭层板，柱面使用多样材质勾勒出丰富的细节，从不同的角度、光线观看都使此场域呈现出独特的故事性。

◇ SPACE PLANNING　空间规划

Designs of the living room and dining room use open design technique. The extension of lines creates the openness and width of the field, from ceiling to TV wall and long sofa, making the whole space produce an internal order so as to create tension of the space. The technique of using angles and curves adds multiple oriented and penetrable senses into the whole space.

客餐厅设计采用开放式的设计手法，透过线条的延伸营造出场域的开阔感与广度，从天花板延续到电视墙，再延伸至长形沙发，让整个空间产生内在的秩序，进而制造出空间的张力。利用角度的雕砌与曲线塑造手法，让整个空间中添增多元面向与穿透感。

◇ LIGHTING DESIGN 照明设计

In the living room, embedded strip lamps and tube lights match with ups and downs of ceiling curvature, dividing the space into various irregular patterns, which creates geometrical aesthetics and a living room with sense of flow.

客厅中内嵌的灯带设计及筒灯配合天花板曲度起伏，将空间切割成多种不规则的图案，搭配组合出几何美感，缔造具有流动感的客厅空间。

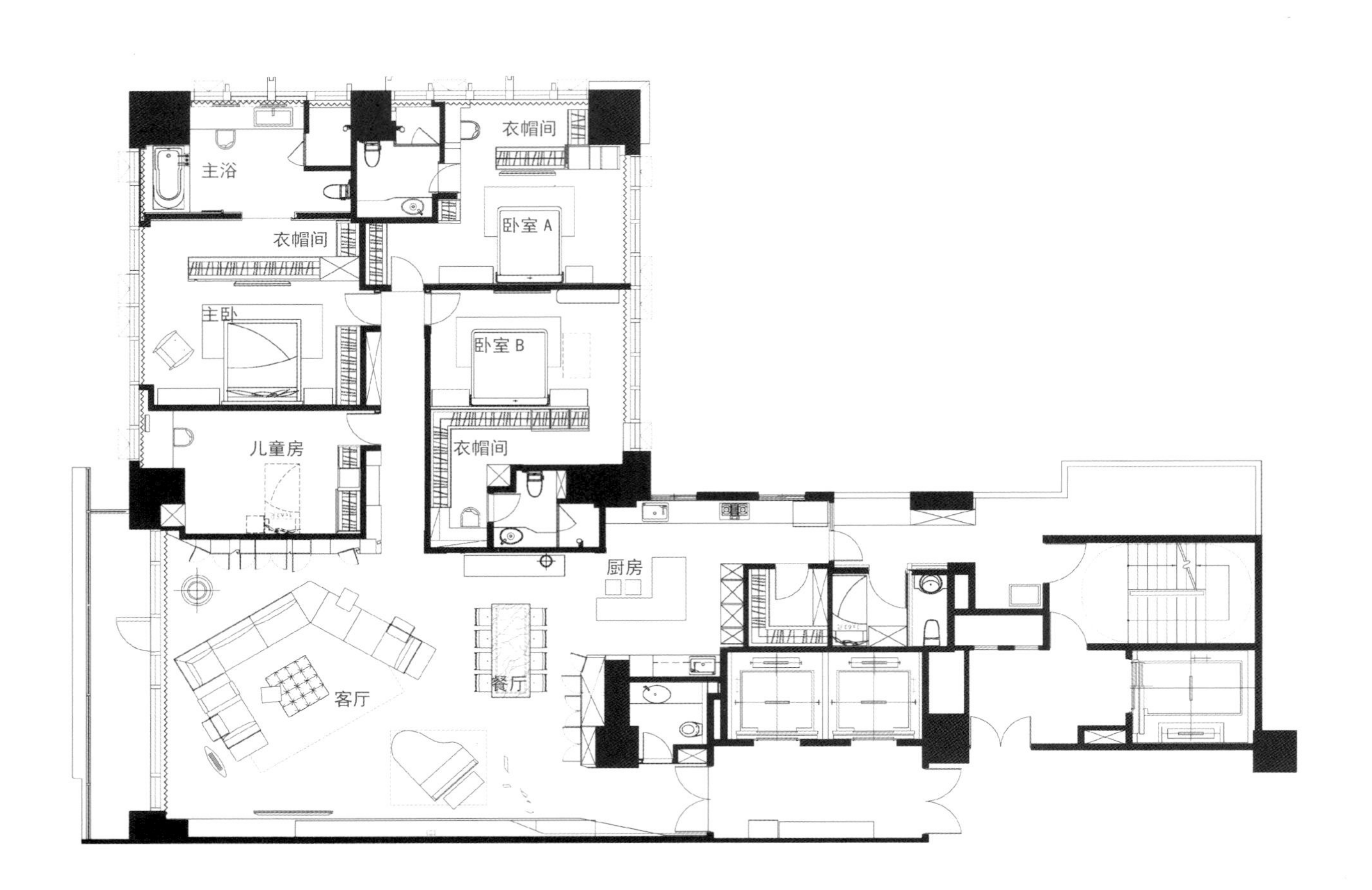

东方风人文宅邸

Oriental and Humanistic Residence

项目名称：久樘香坡
Project name: Chateau de Chambord

设计公司：庄哲涌设计事业有限公司
Design company: Yung Design Group

设 计 师：庄哲涌、陆婉雯
Designers: Chuang Che -Yung, Lu Wan-Wen

项目地点：台湾台中
Location: Taichung, Taiwan

项目面积：215m²
Area: 215m²

主要材料：意大利进口砖、木皮、茶镜、织品、窗帘、山水大理石、比利时木地板等
Main materials: Italian imported tile, wood veneer, tawny glasses, texture, curtain, landscape marble, Belgium wood floor, etc.

摄 影 师：朴叙空间创意有限公司
Photographer: MD PURSUIT

DESIGN CONCEPT
设计理念

The designers give this house unique aesthetic designs, add building materials with expressions according to local conditions, use changes of materials to hint different field properties, introduce restrained and magnificent Oriental style and create a concise and fashionable space charm.

The space interprets white to “dynamic” and wood to “static”, creating an integrated balanced proportion. The contrast colors of the ceiling echo with the floor material, defining different field spirits. Wood are paved in the living room to present the beauty of warmth. The kitchen uses Italian white tiles to show modern and neat tone. Through the multiple use of materials and according to different functional requirements of the field to achieve the convenience of maintenance, the whole row of linear air outlet is arranged in the ceiling, cleverly solving all air conditioning needs of the public area.

设计师给予这个居宅独到的美感设计，因地制宜加入有表情的建材，借由材质的变化隐喻不同场域属性，并导入内敛大气的东方风格，型塑简约时尚的空间韵致。

空间中以白色引申为“动”，木质隐含“静”，拼凑出动静相融的平衡比例，并让天花板的对比色差呼应地坪材质，界定不同场域精神。木地板铺陈于起居空间，借以表现温润之美；餐厨则揉入了意大利白磁砖，呈现现代感利落格调。透过材质的多元使用，因应场域不同机能需求，达到维护的便利性，同时在天花板安排整排线型出风口，巧妙解决了公共区域的所有空调需求。

◇ DECORATIVE MATERIALS 装饰材料

Warm wood material and stiff marble with landscape grains constitute a space form, in addition to the chandelier in pine nut shape. Furnishings which retain Oriental element essences and intentions such as fish group or peony to sprinkle Chinese emotional aesthetics, to express connotative temperament from East and to present noble taste of the owner.

温润的木质材料与山水纹理的坚硬大理石互相构成空间形态，融入松果为型的餐吊灯，保留东方元素精华及意向的饰品，或鱼群，或牡丹，挥洒中式情感美学，表达来自东方的含蓄气质，阐述居住者崇高品味。

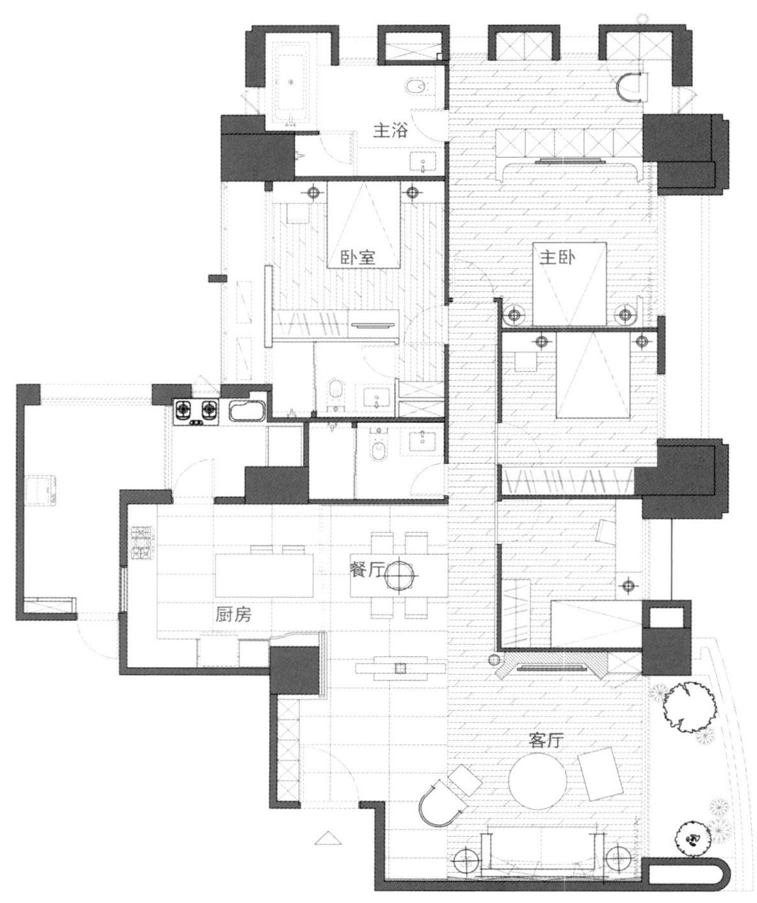

◇ SPACE PLANNING 空间规划

There are no frames between the living room and dining room but they are divided by the hollow-out marble partition which acts as decoration and creates a transparent and deep visual feeling with actual and virtual depth of field. The living room, dining room and kitchen are planned continuously, giving residents a spacious and fluent life feeling.

客餐厅之间并无明显的屋框将两个空间分割开来，而是以中间掏空的大理石隔断作为分界，不仅起到装饰的作用，而且创造通透纵深的视觉感受，有着虚实错落的景深层次。客、餐、厨等空间连续性规划，给予居住者敞朗流通的生活感受。

◇ NATURAL COLORS 自然色彩

The elevation of the living room uses materials such as gray landscape grain stone, dark color wood veneer and floral wallpaper to convert Chinese elements. By cutting and transforming them, a rich visual layering is injected into the space, creating a vivid scene as if the landscape painting. The dark wood color wall and wardrobe in the master bedroom use their plain and texture, magnificent and sedate color and luster to intoxicate people.

客厅立面经由山水灰纹理石材、深色木皮、花卉壁纸等素材，将中式元素加以转化，给予切割、转折等语汇雕塑，注入丰富视觉层次，营造仿若山水画作的生动画面。卧室深木色墙壁及衣柜，朴实无华的质感与大气沉静的色泽令人陶醉。

韵染 Rhythmical Land

设计公司：伊太空间设计事务所
Design company: Etai Space Design Office

设 计 师：张祥镐、高子涵
Designers: Sean, Zihan Gao

项目地点：台湾
Location: Taiwan

项目面积：166m^2
Area: 166m^2

主要材料：铁件、木皮、黑烤玻、绷皮、石材、镜子、玻璃等
Main materials: iron, wood veneer, black burnt glass, fabric, stone, mirror, glass, etc.

摄 影 师：游宏祥摄影工作室
Photographer: Kyle Yu Photo Studio

DESIGN CONCEPT
设计理念

Space has its own rhythm. The designers listen to the rhythm of the exclusive space, use grains of materials to outline of the rhythm of life, create a relaxing and free atmosphere by graceful lines of marbles and render them into every corner of the space to make everyone who comes here to listen to the elegant melody after calming the mind down.

Neat and clean lines structure the space property; delicate and uncomplicated design technique manifests the life tonality of the modern home; the composed and restrained environment presents a humanistic atmosphere; in the foil of new type metropolitan landscape, it reflects an elegant texture residence.

空间都有专属的节奏，倾听空间专属的节奏，运用材质的纹理勾勒出生活的韵律，以大理石曼妙的纹路，创造出轻松、恣意的氛围，渲染到空间每个角落，让来到这空间的每个人在心灵沉静之后再细细聆听感受优雅旋律。

以干净利落的线条架构出空间性质，精致不繁复的设计手法，展现出现代家具生活调性，沉稳、内敛的环境呈现出人文气息，在新型态都会景观的映衬下，反映出优雅的质感寓所。

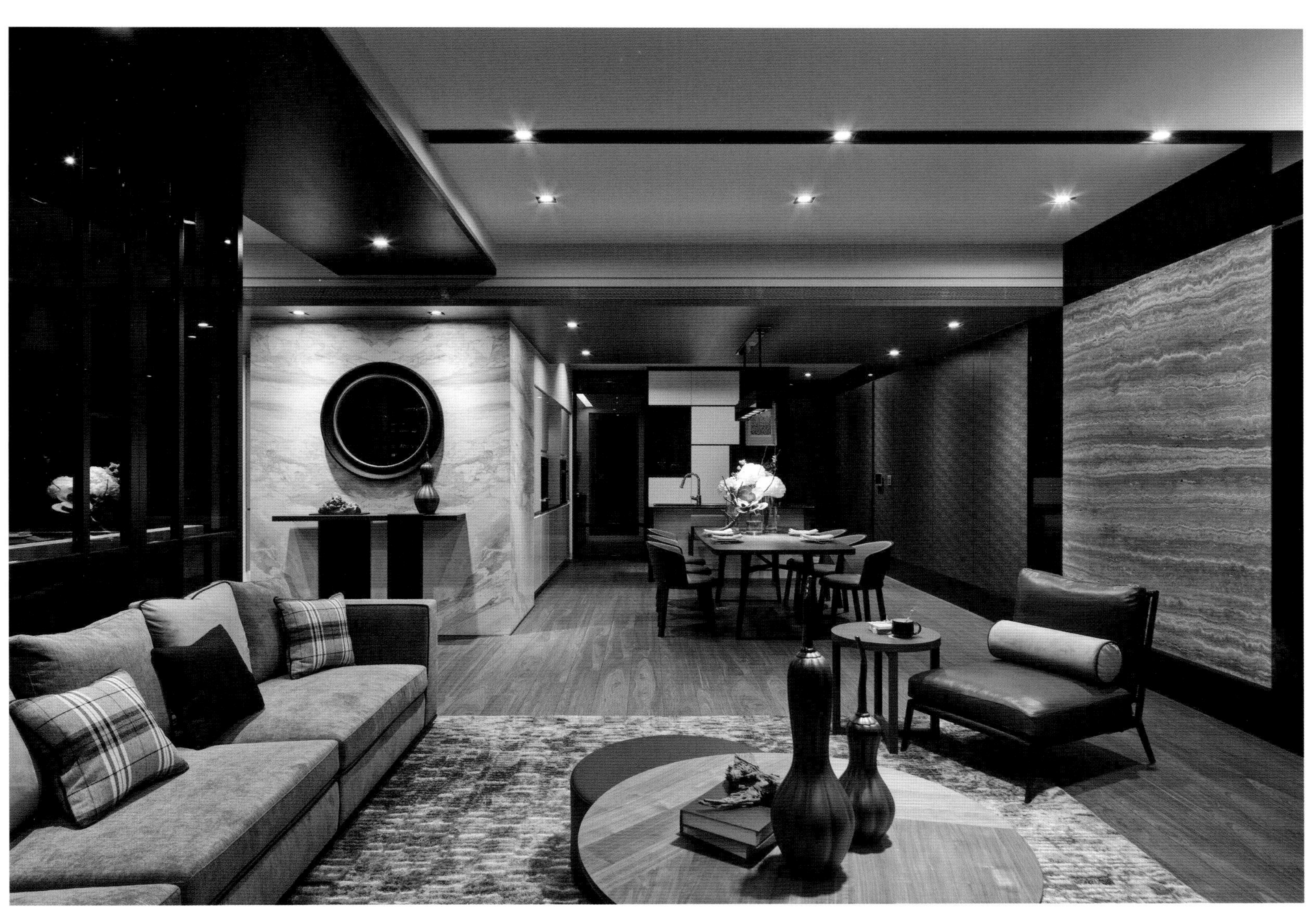

◇ DECORATIVE MATERIALS 装饰材料

The tender wood, the stable stone material and the application of fossil give a variety of expressions for different areas. The exquisite wall made of natural stone material is coated with the iron pieces with the concave and convex. The independent study uses iron pieces to penetrate the wall, making visions communicate in different spaces. The complementary decorations of furniture and software and every detail concentrated in every field show the exquisite atmosphere in the space.

木质的温润、石材的稳定，利用木化石让场域之间的各自有不同的表情，天然石材壁面的细致搭配不过度凹凸的铁件包覆，独立的书房空间，利用铁件的分割穿透墙面，让视觉在不同的空间也能传达交流。家具、软件的辅助搭配，各场域注重着每个细节，让整个空间呈现出精致的氛围。

◇ SPACE PLANNING 空间规划

The lines outline the simple pattern. The application of a wide spectrum of materials adds more texture in the space. The stable colors streaming in the space comfort one's mood. The linked space expands the vision and feeling. The interactive presentation of ceiling, floor and wall builds a humanistic and comfortable space.

线条勾勒出简洁的形态，媒材的运用让空间更具质感，稳定的色泽串流于空间之中，安抚了心绪，场域之间相互连贯开拓了视野及感受，天、地、壁的交互呈现，建构出人文舒适的空间。

◇ LIGHTING DESIGN 照明设计

As for the dialogue of spaces, the private area uses the chandelier to show soft and mellow rhythm, like elegant dancing of the ballet dancer which can make people feel the romantic and comfortable atmosphere. The grain of the stone wall is the dance partner of the chandelier, foils the elegance of the space and creates a comfortable and elegant field where you perceive a modern comfortable humanistic residence.

空间与空间的对话，公私领域借由水晶吊灯导引出柔美的韵律，像芭蕾舞者优雅的舞姿让人感受到空间浪漫舒适的氛围。石材墙面的纹理是水晶吊灯的舞伴，衬托空间的优雅，造就了舒适雅致的场域，感受一个现代摩登的舒适人文住宅。

甲山林 The Best Forest Land

项目名称：水立方B1-40F 实品屋
Project name: Water Cube House B1-40F

设计公司：动象国际室内装修有限公司
Design company: Trendy International Interior Design

设 计 师：谭精忠
Designer: David Tan

参与设计：许婉珍、林怡如
Cooperating designers: Wanzhen Xu, Yiru Lin

项目地点：台湾新北
Location: New Taipei, Taiwan

项目面积：538m^2
Area: 538m^2

主要材料：木皮、不锈钢、灰网石、黑网石、海岛型木地板、皮革裱板、黑水晶马赛克、夹纱玻璃等
Main materials: wood veneer, stainless steel, gray mesh stone, black mesh stone, island type wood floor, leather plate, black crystal mosaic, yarn-wired glass, etc.

DESIGN CONCEPT 设计理念

"Water" which has given birth to both the forest land and mankind is the main concept of the overall design. Water drop falls into the water, sets off ripples, extends the central axis of home and signifies the never-ending relationships amongst family members.

Located near the mangrove MRT station in Tamsui District, this project can overlook the Tamsui estuary and Guanyin Mountain scenery. Its high viewing balcony achieves magnificent and grand landscape mansion and highlights the value and taste life of this project. The design core combines fashion and leisure with art, which breaks the existing layout and maximizes the possibility of the space to make you feel the magnificent manner of large area when entering the interior. Materials such as titanium stainless steel and steel brush dyed wood veneer are interspersed in the space, which manifests the design tension of this project under the restrained connotation of combining with art works.

运用孕育山林大地的“水”为本案设计出发点，水滴落入水面，泛起阵阵涟漪，引申出家的中心主轴，并隐喻家人彼此的关系绵延不绝。

本案位于淡水区红树林捷运站附近，以俯瞰淡水河口、远眺观音山景，挑高错层观景阳台，成就恢弘大器的景观豪宅，突显本案的价值与品味生活。以时尚休闲结合艺术为设计核心，在打破既有格局与发挥空间极大化的可能性，一进入室内便能感受大坪数的恢弘气度。空间中点缀性的融入镀钛不锈钢与钢刷染色木皮等材料，在结合艺术品的内敛涵养下，更彰显本案的设计张力。

◇ DECORATIVE MATERIALS 装饰材料

The water drop specially created by the artist is hanged in the ceiling of ripple modeling and is used to introduce the main axis of this project. The scale of water drop respectively echoes with the family members and is integrated with the five-element symbols which represents life and growth in nature. Both ripple and water drop signify the never-ending relationships amongst family members.

The master bedroom merges another bedroom into it. The arc main wall extends to the ceiling, forming a visual focus. The entire space adopts the consistent comfortable mellow tone; the material applications from bedside leather board, titanium stainless steel to steel brush dyed wood veneer are luxurious, elegant and restrained. The warm tone and Oriental furniture can properly present the owner's taste.

涟漪造型的天花板，悬吊着艺术家特别创作的水滴，借此代入本案主轴。水滴的尺度分别呼应出家庭成员，同时又将代表生生不息的五行象征融入其中，由涟漪至水滴，隐喻家人彼此的关系绵延不绝。

主卧室将原有的另一间卧房并入其中，设计以弧形主墙延伸至天花板形成视觉焦点，空间整体采用一贯舒适的圆润基调，从床头的皮革裱板、镀钛不锈钢到钢刷木皮染色的材质运用，仿如奢华中带点优雅内敛，在温润的色调与东方调性家具的点缀，能适时的表现出主人的品味。

◇ SPACE PLANNING 空间规划

Entering into interior, what you see are the living room, dining room, chess room and open kitchen. The designer skillfully merges the original room into public spaces and changes the closed kitchen into an open one. The large scale open space can better manifest the magnificence of this project.

踏入室内，映入眼帘的是客、餐厅，棋牌室与开放式厨房，巧妙将原有的一房纳入公共空间中，并将封闭式厨房改为开放式厨房，超大尺度的开放空间，更显示出本案的大气非凡。

◇ NATURAL COLORS 自然色彩

The bedrooms are themed with individualized and sedate tonality. The dark color steel brush wood veneer of the porch in the entrance quietly hides the chest of drawers. The partition modeling wall with light color steel brush wood veneer, titanium stainless steel and olive green leather skillfully hides the door of bathroom. Sandbag hanged in the corner brings out the boy's favorite sports. The deliberately lowered head bed modeling soft cushion skillfully separates reading and resting area where you can lean leisurely from sleeping area. The low partition screen in the sleeping functions as bedside table and extends the function of landscape bed, which makes the simple bedroom changeable and interesting and extends vision of the space.

卧房以个性、稳重的调性为主题。入口深色钢刷木皮的玄关，悄悄地将衣柜暗藏于其中。浅色钢刷木皮、镀钛不锈钢与橄榄绿色皮革的分割造型墙，巧妙地隐蔽了浴室的门片。角落垂吊的沙包，带出了男孩喜好的健身运动。刻意降低的床头造型软垫，巧妙地区分了阅读卧榻区与睡眠区，并于卧榻区阅读时可轻松地倚靠。睡眠区的矮隔屏，将床头柜的功能纳入其中，延伸出观景卧榻的功能，使单纯的卧房变得多变有趣，更达到视觉延伸的空间效果。

◇ BRINGING SCENERY INTO HOUSE 引景入室

There pieces of large viewing windows bring the surrounding natural landscape into interior and make it become the most magnificent natural painting. Drinking tea, chatting, reading or appreciating art works, you can enjoy the natural beautiful scenery at the same time. Art works such as *Water Shape Ten*, *The Cosmic Symphony*, *Pine Valley* and *Luanlin Mountain Painting* on the display cabinet naturally integrate indoor and outdoor landscapes into a whole.

三面大型落地观景窗，将周遭自然景观揽进室内，成为最壮丽的自然画作。不论喝茶、谈天甚至阅读、艺术品赏玩，皆可欣赏自然美景。展示柜上的艺术品《水形之 10》《宇宙交响曲》《松壑》《峦林山景图》，自然地使室内外山水景观交融。

层峰汇聚 品鉴时尚尊贵

Peak Convergence, Appreciate Fashion and Dignity

设计公司：成晟室内装修设计工程有限公司
Design company: CHENG SHENG INTERIOR DESIGN

设 计 师：李兆亨、成秋双
Designers: Li Tsan Hen, Cheng Chiu Shuang

项目地点：台湾台南
Location: Tainan, Taiwan

项目面积：442m²
Area: 442m²

主要材料：木皮、大理石、镀钛、铁件等
Main materials: wood veneer, marble, titanium, iron, etc.

摄 影 师：郭家和
Photographer: Ar-Her Kuo

DESIGN CONCEPT 设计理念

This project is located in the downtown area of Tainan, with an area of more than 300 square meters after opening up two houses. The designers use straightforward and casual style to create a magnificent and spacious mansion. Based on excellent conditions of the project such as “spaciousness” and “bathing in lights” near the sea, the designers use environmental awareness to intersect visual design ideas in the design style, regard every space as a unique stage and create the essence of home to be the main axis of designs.

The true essence of the home is human, composition of emotions connects the precious times among family members. The host who often does business at home and abroad wants a tranquil life. Therefore, the designers create a modern and leisure living life, integrate exterior environment into the home design, combine the spaciousness of exterior environment with modern straightforward style, create a leisure style and express the emotional atmosphere of the whole space.

本案位于台南市区，两户打通超过 300 平方米的大空间，以率性的休闲风格，打造气派宽敞的大户空间。案例的“阔”及临海“沐光”的优良条件，运用环境意识在设计风格中穿梭视觉设计观点，将每个空间都视为独一无二的演绎舞台，创造出符合家的本质为此次设计的主轴。

真正构成家的本质，其实是人，情感与情感的构成，更凝聚了家人间的流经岁月。经常往返国内外经商的男主人因此想塑造出一种宁静的生活，取其特性，打造现代休闲家居生活，将外部环境融入居家设计中，与外部环境的气阔结合现代的率性风格，营造出休闲风格，表述了整体空间的情感氛围。

◇ DECORATIVE MATERIALS 装饰材料

The designers use the theme of "creating a fashionable and exalted mansion" to present the unique space aesthetics in the field and pay attention to the uniqueness of materials. For example, TV main wall is constituted by stones with natural texture while the elevation and the side view are paved with warm wood veneer.

设计师以"打造时尚尊贵宅邸"为主题，来展现场域内特有的空间美学，并讲究材质的使用独特性，如电视主墙采用天然肌理的石材所构组，立面及端景部分，则以温润感的木皮铺设。

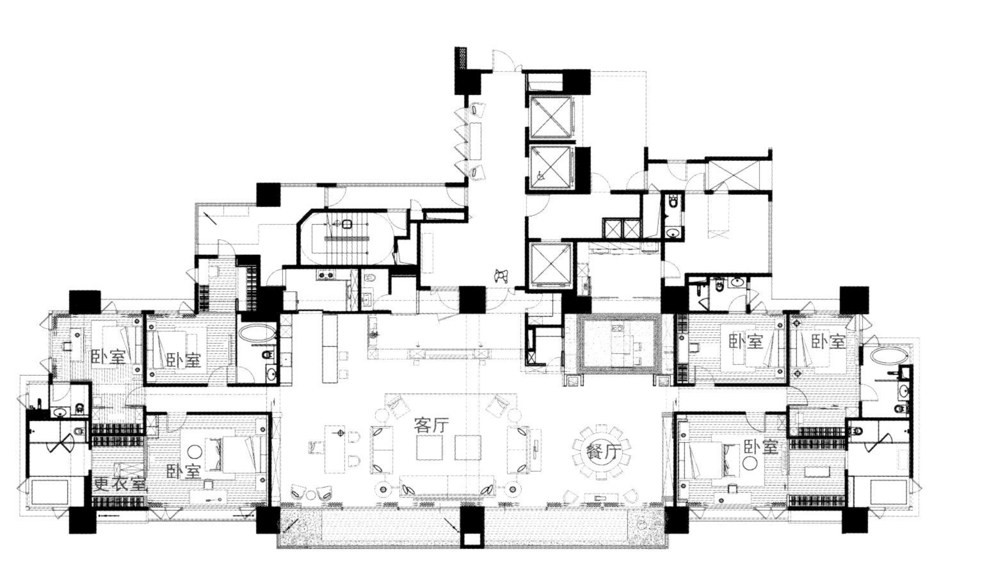

◇ SPACE PLANNING 空间规划

Balance is not only the philosophy of life, but also the indispensable modeling principle of design. The living room, dining room and study with a area of about 200 square meters after opening up the two houses use no compartment to present a mansion manner. Because the living room is the intermediary point of the two houses, the ceiling is specially designed to maintain the consistency of horizontal lines of them. Customized sofa furniture is set in symmetrical way to extend "expansion", to strengthen the integral style and to unify the visual balance.

平衡不只是人生哲学，更是设计中不可或缺的造型原理。在两户打通约 200 平方米的客厅、餐厅、书房，以无隔间方式展现大户气派。由于客厅是两户的中介点，因此在天花板上，特意做出设计，来维持两户水平线的一致性。特制的沙发家具，以对称的方式摆放设计、延伸"阔张"，加强风格整体性，并让视觉平衡统一。

◇ BRINGING SCENERY INTO HOUSE 引景入室

Every bedroom from the the living room to the west side has extraordinary sea view. Quietly enjoying the sparkles of sunshine, bathing in the home design, lying down on the chair or sitting near the bed, you can look at the vast waters, which can wash all fidgety matters. The mind is open and the heart smiles. Home, is a charging station of human; home, is the point of life; home, is the best haven.

包含客厅到西侧的每个卧房，都坐拥无敌海景，宁静享受着阳光的溢洒，浸沐在居家设计里。或是躺卧在靠椅上，或与坐卧在床边，望向一面辽阔的海域，真能一洗烦躁的俗事。不但心境开了，笑容更上了心头。家，是人的充电站；家，是安身立命的所在点；家，是最佳的避风港。

低调奢华 绝世品味

Low-Profile Luxury and Unique Taste

设计公司：成晟室内装修设计工程有限公司
Design company: CHENG SHENG INTERIOR DESIGN

设 计 师：李兆亨、成秋双
Designers: Li Tsan Hen, Cheng Chiu Shuang

项目地点：台湾台南
Location: Tainan, Taiwan

项目面积：248m²
Area: 248m²

主要材料：铜刷栓木、石材、铁件、镀钛、订制家具等
Main materials: copper brush bolt wood, stone, iron, titanium, customized furniture, etc.

摄 影 师：郭家和
Photographer: Ar-Her Kuo

DESIGN CONCEPT
设计理念

The designers choose leather and wood, natural and warm wood furniture to present the warmth of the home and a perfect living atmosphere. Restrained and elegant leather texture with concise and neat arc cuttings foils the extraordinary living breath. Modern adornments inject the space with noticeable fashion charms.

Cheng Sheng Interior Design uses modern design languages and earth tone to interpret mansion model with unique taste, in combination with the arrangement of indoor lighting and elegant furnishings, creating a grand and magnificent, warm and touching aesthetic atmosphere.

设计师挑选皮革、木质等异类质材作为选搭，自然且充满温暖触感的木作家具，将居室温润而居家的氛围完美构造。内敛高雅的皮革质感，伴以简约利落的弧形剪裁，则烘托出质感非凡的家居气息。而现代感十足的现代装饰品，则为空间注入不容忽视的时尚风采。

成晟设计透过现代感的设计语汇，以大地色彩诠释出品味出众的豪邸典范，结合室内光源的安排，以及典雅风情的家具家饰，在轩昂大气之中又形塑一股温润且动人的美学氛围。

◇ DECORATIVE MATERIALS AND SPACE PLANNING 装饰材料、空间规划

In the living room, the ground is covered with plain and elegant carpet, collocated with leather sofa and chairs which are carefully chosen, foiling the extraordinary and exalted temperament of the owner. Behind the sofa, there is a study with strong color wood grain cabinet, making the reading space vitalize with nature; with the combination of mirror design, its perspective characteristics make the space more extensive and spacious; in addition to storage and display functions, it meets the needs of visual aesthetics and practical features. The ceiling under the dining table uses arc images to outline the visual modeling; the below dark color plain and elegant dining table means reunion in Oriental kindness; the behind cupboard with gorgeous porcelains adopts mirror basement to add luxurious texture along with the reflection of lights.

客厅里以素雅的地毯铺陈地坪，并严选皮革沙发与座椅且相互搭配，烘托屋主自身非凡且尊贵的气质；沙发的后方规划出书房区，佐以木纹色彩浓厚的柜体，让阅读空间伴随自然生气，而透过镜面设计的结合，其透视特性让空间更为延展、宽阔，加上收纳与陈列机能兼备，一次满足视觉美感与实用特质。餐桌区的天花则以圆弧意象勾勒视觉造型，下方为素雅的深色餐桌，隐喻团圆的东方美意，而后方餐柜摆设华美的餐瓷，并以镜面为底，随着光源反射更增添奢华质感。

◇ NATURAL LIGHTING 自然采光

Through large-scale of window scene arrangement, the designers not only bring a lot of outdoor lights into the living room, but also introduces natural green from the balcony plants, creating a comfortable and leisure living atmosphere. Bedrooms also need lights; both sides of windows are decorated with soft curtains, transferring natural lights into tender lights so as to create a comfortable sleeping atmosphere.

设计师透过大开的窗景安排，不仅户外大片阳光大片倾泻入客厅，而且引入阳台处植栽的自然绿意，家居氛围转以惬意悠闲。卧室同样需要阳光浸润，两侧窗户饰以柔美窗帘作为搭配，让自然光进入室内转换为温柔光感，营造舒心的睡眠氛围。

绿意生活 Green Life

项目名称：靓海筑
Project name: Beautiful Sea-View Residence

设计公司：橙田建筑｜室研所
Design company: CHAIN10 URBAN SPACE DESIGN

设 计 师：罗耕甫
Designer: KENG-FU LO

项目地点：台湾高雄
Location: Kaohsiung, Taiwan

项目面积：1520m^2
Area: 1520m^2

主要材料：清水模、大理石、白橡木、胡桃木皮、LO-E玻璃、铁件等
Main materials: clear water mould, marble, white oak wood, walnut wood veneer, LO-E glass, iron, etc.

摄 影 师：李国民
Photographer: Kuomin Lee

DESIGN CONCEPT
设计理念

The building is located in Kaohsiung in southern Taiwan, faces the unloading zone of Kaohsiung harbor with sea wind breezing from southeast throughout the whole space and with perfect viewing conditions. The house is used as office and residential space. Apart from entrance and garage in the first floor, the designer raises the height of floorslab and uses water-view flower-stand as compartment of the entrance; floors above the third floor are shortened inward for the convenience of the owner. In order to create a life space connecting with nature, with several viewing balconies, the designer uses lots of energy-saving French windows besides the natural wood grain clear water mould of the building, expecting that the beautiful harbor scenery can be seen at any time.

The designer uses the indoor top as living room, dining room and kitchen, expecting to combine with outdoor terrace and the half-open indoor public space to bring the family better life interactions. The interior designs abandon excessive decorations, only highlight the users' storage practicality. What is left is the white space of interior which corresponds to the outside nature, making natural wind blow and stay and making space talk with time, wind, air flow and shaking shadows of leaves.

这栋建筑物位处台湾南端高雄市，面对高雄港码头卸货区，东南向的海风吹拂贯穿整体空间外也有极佳的观景条件。业主作为事务所及住宅空间复合使用，一层楼除了住家出入口与车库用地外，设计师更抬高楼板高度及外覆水景花台以作为出入口的区隔，三层楼以上内缩楼地板作为住家使用。为了创造一个与大自然连接的生活空间，搭配多处观景阳台，设计师在建筑物外覆自然体态的木纹清水模外，亦大量采用了节能的落地玻璃窗，期望美好的港阜风景能够随时映入眼帘。

设计师将室内顶层留作为客餐厅及厨房使用，期望结合户外露台及半开放的室内公共空间能够为家人带来生活上最好的互动。室内设计不做过度的装饰，仅考虑到使用者的收纳实用性，留下的是与室外大自然对应的空间留白，让自然的风吹拂留驻，让空间可以跟时间对话，跟风对话，跟气流对话，跟抖动的树间叶影对话。

按摩室
主卧室
更衣室
阳台
浴室

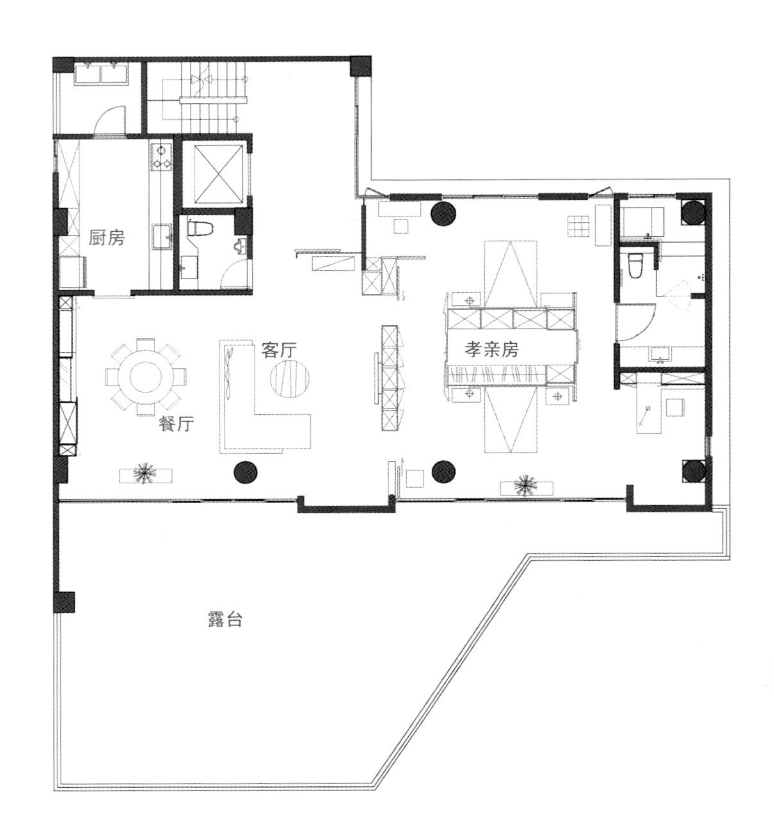
厨房
客厅
孝亲房
餐厅
露台

◇ SPACE PLANNING AND DECORATIVE MATERIALS 空间规划、装饰材料

厨房
礼佛堂
客厅
餐厅
阳台
阳台

The designer lines out the living and dining area in the porch, wine cabinet and straight clear water mould wall. The dining table in the dining area and island vertically correspond to each other and create a lively space. Behind the island, the low cabinet is set with all kinds of drinks to show a low key bar culture; the high cabinet on the right side is placed with oven and hidden refrigerator, making a practical working area. Walnut solid wood table is placed on the right side of the island, enhancing interactivity of kitchen utensils.

The open living and dining rooms and kitchen area match with the balcony and the nearby harbor sea view, creating a perfect rest life environment. From the elevator exit, you can see branches of beech and green maple trees swaying with the wind. With time differences, the attic skylight daylighting flows in the clear water mould wall, leaving dynamic changes of light and shadows. The designer makes open and half open balconies form a funny dialogue. The light and shadow from the top of the tree freely change and wander in the stone floor and white walls as time goes by.

设计师在室内空间的玄关区、酒柜及笔直清水模墙区划了客餐厅区，餐厅区的餐桌与中岛台垂直对应创造出空间的活泼性，岛台后侧低柜各式酒品随兴陈列，展现出低调酒吧文化，右侧高置物柜的蒸烤炉及隐藏式冰箱串连了工作区的实用性，胡桃实木横亘在岛台上方的右侧，提升了厨具使用上的互动性。

开放式的客餐厅、厨房区搭配露台及彼临的港湾海景，营造了绝佳的休憩生活环境。电梯出口可以看见榉木及青枫树梢末端随风摇曳，随着时间差，顶楼天窗采光流泄在清水模墙留下动态的光影变化，设计师让开放及半开放的阳台空间形成趣味的对谈。从树梢洒落的光与影随着时间的流逝，尽情地游走在铺满级配料石片的地面与洁白的墙面。

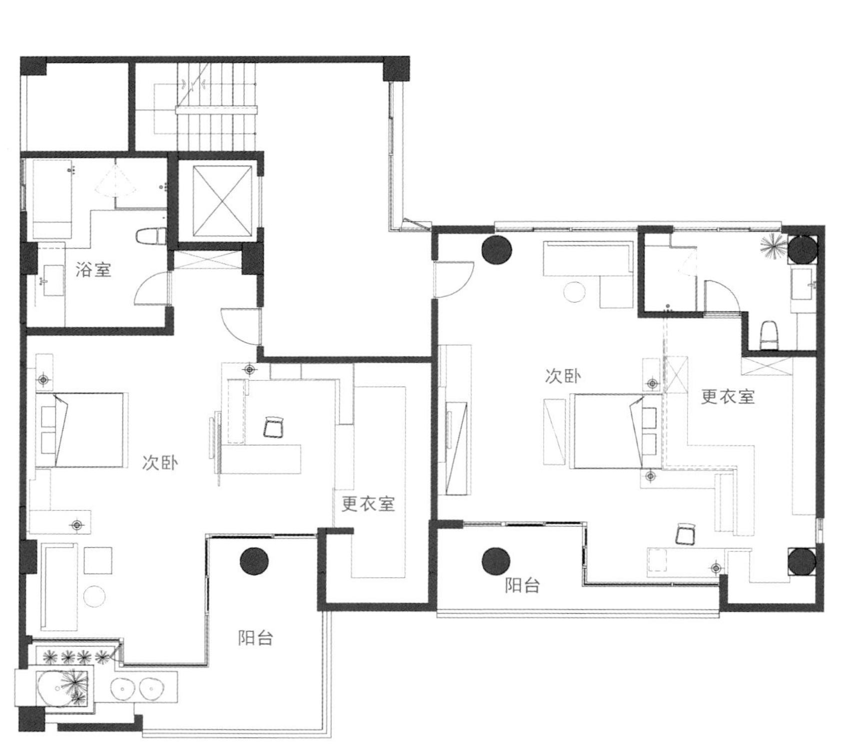
浴室
次卧
更衣室
阳台
次卧
更衣室
阳台

◇ LIGHTING DESIGN 照明设计

The indoor lights sparkle over the stainless steel island and reflect to the walnut ceiling, shining warmth of the whole room, as if declaring the play of the main character. Many hidden tube lamps which are embedded inside the ceiling flicker add the soft atmosphere for the space, bringing a warm and comfortable feeling.

室内灯光投射在中岛台不锈钢台面反射到胡桃木天花板，辉映了满室的温暖，似乎在宣示空间中的主角演出。而嵌于天花板内部的多盏隐藏筒灯闪烁，增加空间的柔和气氛，带来温馨自在的感觉。

熙·境 Light and Place

设计公司：杨焕生设计事业有限公司
Design company: YHS Studio

设 计 师：杨焕生、郭士豪
Designers: Jacksam, Steven

项目地点：台湾彰化
Location: Changhua, Taiwan

项目面积：248m²
Area: 248m²

主要材料：木皮、钛金属、订制家具、订制灯、镜面、大理石等
Main materials: wood veneer, titanium, customized furniture, customized lamp, mirror, marble, etc.

摄 影 师：岑修贤
Photographer: Xiuxian Cen

DESIGN CONCEPT 设计理念

The designers think, design is not to divide the space changelessly, but to inject featured ideas, memory points and natural elements into each space so as to create a dramatic space.

In this project, the designers abandon luxury and complication and use the simplest lines to outline the manner and connotation of the space. Based on black, white and gray tones, elegant and sophisticated images are presented. The designers carefully select building materials and use elegance to create a delicate luxury. The house is not only the extreme convey of aesthetic feeling, but also an expression of a sweet home. "Xi", the light, means brightness and prosperity. The designers use dividing lines, tones and materials in the space to create a harmonious balance between calmness and transparent white, making the house a bright place as if in the land of idyllic beauty.

在设计师的眼中，设计并不是一成不变的分划空间，而是将特性的想法、耐人寻味的记忆点、灌溉自然原素注入到每个空间中，才能谱写出扣人心弦的空间乐章。

本案中，设计师减去华丽繁复，用最简单不过的线条，勾勒出空间的气度和内涵；用基础的黑白灰色调，描绘出优雅且精致的意象；精选建材，用雅致装饰一份细腻的奢华。居室不仅是美感的极致传达，也是温馨的家的所在。"熙"，光也，是光明，是兴盛。设计师借以空间中分割的线条、以色调、以材质的呈现，在沉稳与透白之间搭配和谐的平衡，将居室打造成如世外桃源明亮中的境地。

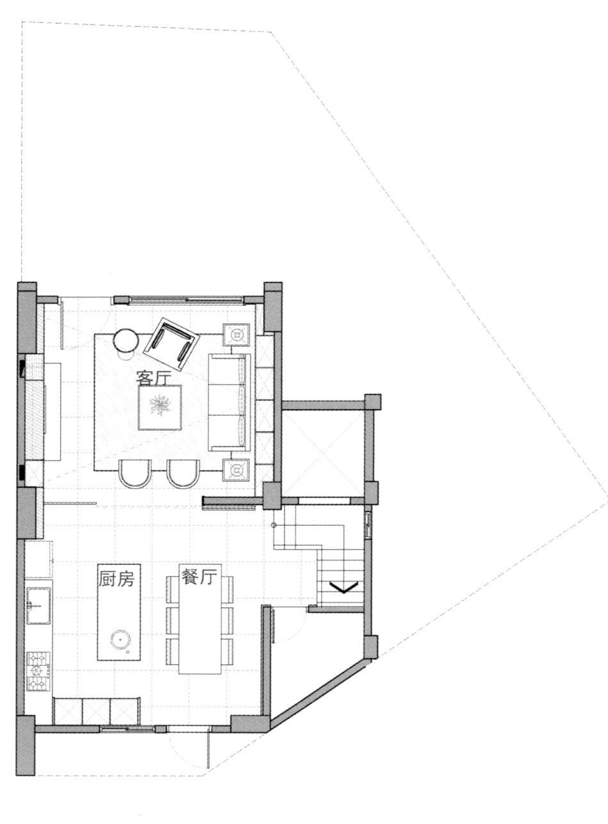

客厅
厨房
餐厅

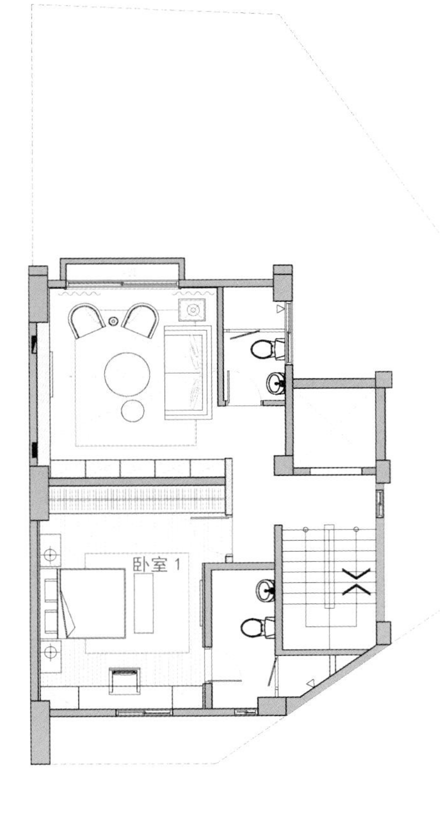

◇ DECORATIVE MATERIALS 装饰材料

The integration of wood veneer and mirror material creates a calm, restrained and fashionable atmosphere. The floor and TV background wall in the living room are paved with dark and light white grain marbles, powerful and magnificent. The dark deduces magnificent manner while the light enlarges the visual effect and creates an exquisite luxurious feeling.

木皮和镜面材质的交错，混搭出沉稳内敛又具时尚感的氛围。客厅地板及电视背景墙用洁白富浅色纹路的大理石大片铺陈，气势磅礴。深色的演绎大气风范，浅色的则有放大视觉效果，创造细腻的奢华感。

◇ SPACE PLANNING 空间规划

Dark color grilling separates the living room from dining room, clearly defining the boundary while not blocking the sight. Its existence is obscure, which enables the sight to pass through and add fun, faintly transforms space functions and makes life become comfortable at will.

深色格栅把客厅与餐厅隔开来，清楚界定却又不阻碍视线穿越，它的存在若有似无，可以让视线穿梭迂回且别富生趣，于隐约间转变空间机能，生活变得自在而随心所欲。

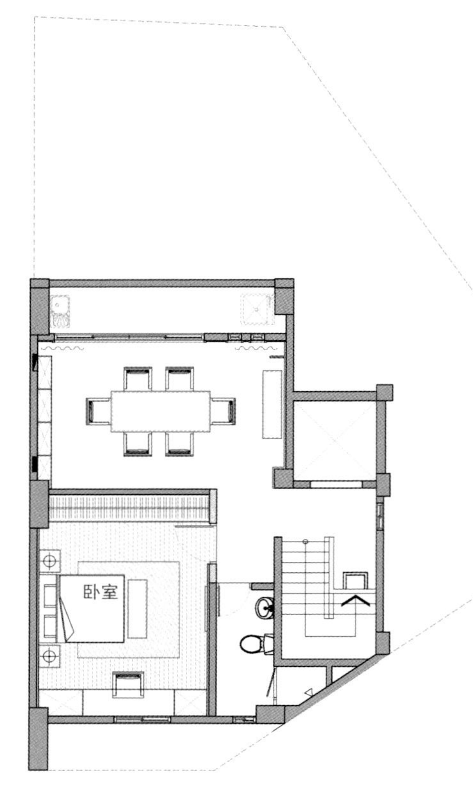

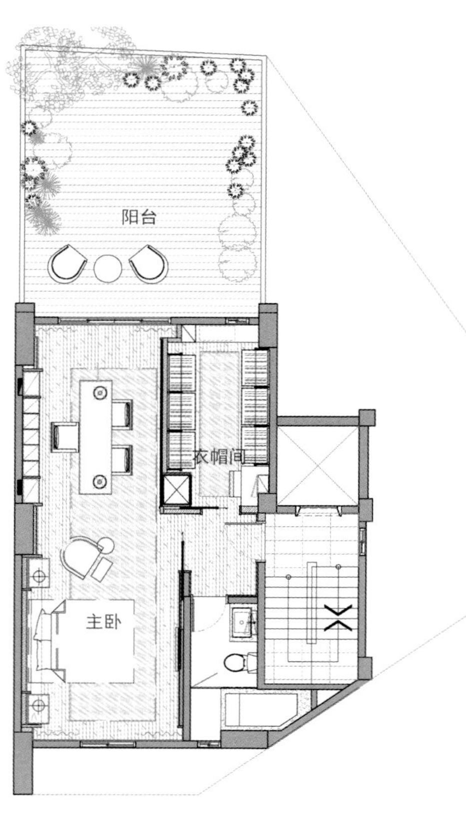

◇ NATURAL COLORS 自然色彩

The house is based on black, white and gray tones. Although it is constituted of simple tones, it can present infinite images. Dark or light, overlapping or scattered, it has clear levels. The dark wood color floor of the rest room collocates with exquisite senior gray cloth of the bedside, restrained and charming, making the bedroom tranquil and comfortable.

居室以黑白灰作为基础色调，虽然只是简单的色调组成，亦可变幻出无尽意象。或深或浅，更迭交错，层次分明。休憩空间的深木色地板搭配着床头精致的高级灰表布，内敛且深含韵味，让卧室空间宁静且惬意。

设计公司：大雄设计
Design company: Snuper Design

设 计 师：林政纬
Designer: Cheng Wei Lin

项目面积：183m²
Area: 183m²

主要材料：普罗旺斯大理石、黑网石材、咖啡绒石材、石材马赛克、超耐磨木地板、铁件、灰玻、铁刀木皮、特殊漆、绷布等
Main materials: Provence marble, black mesh stone, coffee velvet stone, stone mosaic, Super durable wood floor, iron, gray glass, Cassia siamea wood veneer, special paint, fabric, etc.

摄 影 师：李国民
Photographer: Kuomin Lee

DESIGN CONCEPT
设计理念

The designer uses modern design ideas to deduce classical feelings; modern complex lines and rich levels deduce the classical complicated concept in some degree. It is neither Oriental nor Zen-like, while it injects sensitive temperature into rational lines, soft tones and furnishings into the stouthearted pattern to refine such a “unique” and “incongruous” space design.

In this mountain residence with spring smell, associating by means of art, dancing and singing, the guests and owners are thoroughly enjoying themselves. This private club here receives important guests in a clam and low-key way, which also represents the hospitality of the host. The meanings don’t need too many words.

设计师运用现代的设计思路来演绎古典情怀：通过现代的复杂线条和丰富的层次，某种程度演绎出古典繁复的概念。并非东方，亦非禅意，而是让理性的线条包纳了感性的温度，在刚毅格局中加入了软性的色调、陈设，提炼出如此“独特”、“违和”的空间设计。

山居春意，以艺会友，轻歌曼舞，宾主尽欢。这处私人会所在山居春意间，沈稳低调地接待了重要宾客，也代表着主人的待客之道，意在不言中。

◇ DECORATIVE MATERIALS 装饰材料

Natural design vocabularies make art works high-spirited and vigorous and shine glories in their own corner. Whether it is the background wall based on stone grains, wall cabinet covered by wood or hand-painted art wall, all is exquisite and low-key, and makes undeniable contributions for the setting of exhibits.

自然的设计语汇，让艺术品更为意气风发，在属于自己的角落中绽放光彩。无论是以石材纹路为基底的背墙、让木质包覆住的壁柜，甚至是手工涂抹的画作墙，细腻却低调，作为展品的背景，功不可没。

◇ SPACE PLANNING 空间规划

With the private club concept, the space planning integrates functions including art gallery, dancing floor and vacation club. There are various recreational activities for distinguished guests who can even live here for three nights and feel relaxed and comfortable. The designer covers wood floors in the long corridor behind the living room; the front is planed into a mahjong room and the middle a dancing floor, providing spacious recreational spaces. The place near the dancing floor sets a fog glass sliding door and elevates the wood floor to define the perfect social space. The corridor can be used as a dancing floor and mahjong room, and the social area can also be a rest area, where guests can spend the night.

空间规划以私人会所概念，融合了艺廊、舞池、度假会所等使用功能，尊荣的宾客享有多种娱乐活动，即便是歇住三宿也能轻松自在。设计师运用客厅后方的长廊，铺陈木质地坪，前段规划为麻将间，中段作为舞池空间，提供宽敞的娱乐空间。舞池旁以雾玻璃拉门，并架高木质地坪，区隔出交谊区，带来完善的社交空间。长廊可作为舞池与麻将间，交谊区也可成为休息区，提供宾客打通铺夜宿的地方。

◇ LIGHTING DESIGN 照明设计

Under the shining floor-lamps, the gallery is no longer just a transitional space, instead it can be a social area for dining and party or a dancing floor for people to dance. The floor-lamps are lit up along the arc sequence, shining light and shadow when people are dancing and singing, which is beautiful and charming.

在情境地灯的营造下，展廊不再只是过渡空间，可以是宴客用餐的交谊区，亦可是让人翩翩起舞的舞池。地灯沿着弧线序列点亮，轻歌曼舞之际，光影闪熠，美丽动人。

沉静湛蓝 与日光漫谈

Quiet and Azure Blue, Talk with Sunshine

设计公司：大漾帝国际室内装修
Design company: Dayoungdi.Design

设 计 师：戴铭泉
Designer: Tai Ming Chuan

项目地点：台湾台北
Location: Taipei, Taiwan

项目面积：125m²
Area: 125m²

主要材料：大理石、海岛型木地板、进口壁纸、进口瓷砖、进口花砖、玻璃、天然木皮、表布、画师画作等
Main materials: marble, island type wood floor, imported wall paper, imported tile, glass, natural wood veneer, table cloth, painting, etc.

摄 影 师：Hey!Cheese
Photographer: Hey!Cheese

DESIGN CONCEPT
设计理念

Quiet and bright visual image is described in the field. The designer uses blue as the basic color, and matches it with other heavy and light colors. The application of main wall, sofa, cushion and carpet brings tranquil and sedate atmosphere into interior. The white marble and tile with carvings quietly render elegant scenery like landscape splash-ink in the glittering and translucent white elevation through black and white dark and light grains.

In creating the style, the designer injects contemporary fashionable vocabularies into the clean and pure lines and surfaces under the frame of neo-classical style, abandons complicated lines and decorative elements and uses concise line boards and art works to present delicate aesthetics. Of course, the function set is also the key point of this project. Every area is set with enough collecting cabinets. No matter it is laminate, drawer or flip-up form, all can perfectly meet the living needs of the family.

将沉静明丽的视觉意象，铺述在场域之中。设计师以湛蓝为底，绘上浓淡有致的色彩，透过主墙面、沙发、坐垫与地毯等运用，将宁静沉稳的氛围带入室内。并搭配雕刻白大理石与磁砖，透过黑白深浅的纹理，在晶莹洁白的立面上，轻轻点染山水泼墨般的雅致风景。

风格形塑上，大漾帝国际室内装修在新古典的架构下，挹注当代时尚的语汇，于干净且纯粹的线面之中，舍去繁琐线条与装饰元素，用简约线板及艺术作品体现细腻美感。当然，机能设置也是本案的重点，每一区均设置充足的收纳柜体，不论用层板、抽屉或上掀等形式，都能完美符合屋主一家的生活需求。

◇ DECORATIVE MATERIALS 装饰材料

The porch uses abstract painting to add solemn and tranquil atmosphere. Luxurious and nifty retro tiles deduce the appearance of neo-classical style starting from the porch. After crossing the porch, it is paved with Caribbean blue and magnificent stones. Transparent sun lights penetrate into interior through the white shutters, lighting up the whole living room. TV wall uses frame stones to add three-dimensional sense and layering sense to the space. The dark color ground stone increases calmness, making the finishing point.

玄关以抽象画增添庄严静谧的氛围，华丽俏皮的复古砖从玄关一路拼接演绎出新古典的风貌。穿越玄关后，是加勒比海的蔚蓝与气势十足的石材。通透的阳光透过白色百叶窗穿透进室内，打亮整个客厅。电视墙采用框面石材，替空间增加立体感及层次感，而深色的地面石材增加沉稳度，有画龙点睛的效果。

◇ SPACE PLANNING 空间规划

The highlights of the space planning are the skillful arrangement and hide of collecting space. The study and living room are in the same axis; there is a desk elevation behind the TV wall with magnetic white paint, which is convenient for owner to note life matters. Chic light decorations in the dining room and living room share the same style; the elastic dining table is a magic which can be pulled out easily.

空间规划的亮点在于收纳空间的巧思安排及隐藏。书房与客厅构筑为同一条轴线，并于电视墙后方的书桌立面，设置具有磁性的优白烤漆，方便屋主随时备注生活事项。餐厅别致的灯饰与客厅灯饰为同系列，可伸缩式的餐桌，仿佛空间魔法般，只要轻松拉出即可往外延伸。

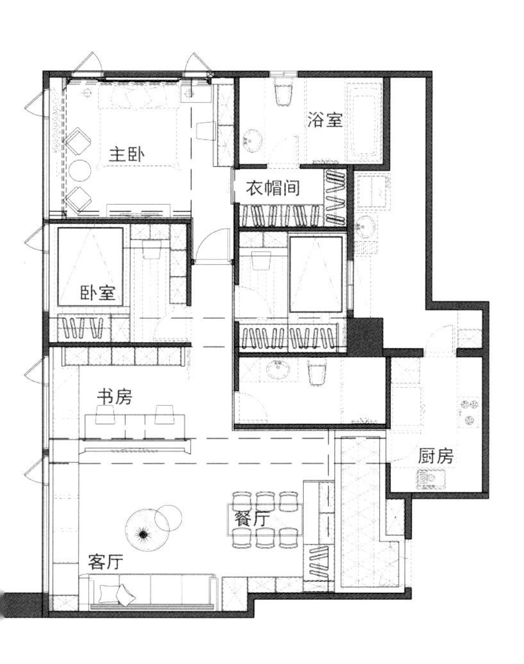

◇ NATURAL COLORS 自然色彩

The owner's favorite blue sets tone for the space. Rich blue tones create a tranquil beauty. Elegant blue, warm blue and calm blue in various appearance are patiently coated to create abstract and aesthetic blue art. Low-key and gorgeous azure blue perfectly combines with elegant silver fox marble, creating a gorgeous, magnificent and elegant tone spontaneously.

将屋主喜爱的蓝为空间定调，丰富的蓝色调创造出宁静的美感况味。优雅的蓝，温暖的蓝，沉静的蓝，混合各种样貌的蓝色，耐心地经历多次细工涂装，创造出抽象美感的蓝色艺术。低调绚丽的湛蓝，与高雅银狐大理石的完美结合，一种华美大气亦不失优雅的情调油然而生。

Offices

礼兰 Li Lan

设计公司：杨焕生设计事业有限公司
Design company: YHS Studio

设 计 师：杨焕生、郭士豪
Designers: Jacksam, Steven

项目地点：台湾台北
Location: Taipei, Taiwan

项目面积：92m^2
Area: 92m^2

主要材料：大理石、磁砖、镜面、茶玻、皮革、钢琴烤漆、订制家具灯具等
Main materials: marble, tile, mirror, tawny glasses, leather, piano coating, customized furniture and lamp, etc.

摄 影 师：林明福
Photographer: Mingfu Lin

DESIGN CONCEPT
设计理念

The starting point of the design concept is from the two Chinese characters ‘Li’ and ‘Lan’, meaning the appropriate behaviors and the delicacy and elegance. Seeking a livable residence in the metropolis is what the residents want for a long time.

Being tired and passing through the busy and prosperous street, one can come to the old house in the heart of the city. Opening the gate, the porch immediately lights up a soft and beautiful atmosphere to welcome the owner’s return, giving people the closest feeling of belonging. Walking into interior and being in the field with British classical breath, the overall harmonious tone and materials create a graceful and elegant space appearance, like a top resort hotel, giving people leisure, relaxing and elegant texture and taste.

设计以“礼兰”为思考起点，“礼”意喻家人之间合宜的行为，“兰”是清秀优雅的象征。在都市中寻得一处合宜的居住场所，是居者一直以来的心之所向。

披戴着一身疲惫，穿越车水马龙的繁华街头，来到藏身于精华地段的老旧住宅。旋开大门，玄关立刻点亮欢迎主人归来的柔美光氛，让人一进门就感受到最亲近的归属感。移步入室，置身充满英式古典气息的场域内，整体和谐的色调与材质运用，交织出雍容高雅的空间面貌，有如顶级度假酒店般，予人慵懒、放松又不失优雅的质感与品味。

◇ DECORATIVE MATERIALS AND NATURAL COLORS 装饰材料、自然色彩

Once the context of the space is clear, the concept of 'ink spread' implies within the design. As seen throughout the floors and walls, gray stone textures with sharpened appearance endow the space with concise personality; the laying patterns integrate phases of floors and walls into a whole. Furthermore, the touch of furnishings adds up a progressive layering of the space. On the vertical wall, a protruding short scaled facade uses different laying and changes in materials to deliver the softness and layers of the space, bringing the 'ink spread' concept; the place, the scenery and objects together, integrates as a whole, making the space present itself in its unique and refined way.

厘整好空间条件后，设计师随之灌注"墨晕"概念，于地板及墙面运用大量灰色石材分割，塑造空间简洁个性，渲染般的层次将空间地壁融合为一体。随之在与家具陈设融揉交衬，在空间构图中迭出远近层次感，垂直墙面运用进出面的变化，将短屏风造型墙面，运用分割及材质变换带出空间的柔软及层次，汇合"墨晕"概念，境、景、物，合宜融为一体，空间如青秀兰花一样品味自出。

◇ SPACE PLANNING　空间规划

Nestled in the heart of city, this inhabitable site was initially narrowed and cramped. Transformed from redesigning the layout and re-purposing a few elements, results in the outcome of a well balanced interior, that brings out a clean-lined scale. At the same time, the designers remove the wall of the former kitchen and set up a central island to make an open kitchen area, which promotes the interaction between the family and makes home full of flavors and warmth.

原先的场域偏窄且格局狭小，经由设计师调整墙面空间及将使用行为稍做改变，调整合适水平景深及纵宽，酝酿出简洁的线条尺度。同时拆除原先厨房的隔间墙，加设一座中岛打造开放式餐厨区，增进家人间的互动关系，让家更有味道和温暖。

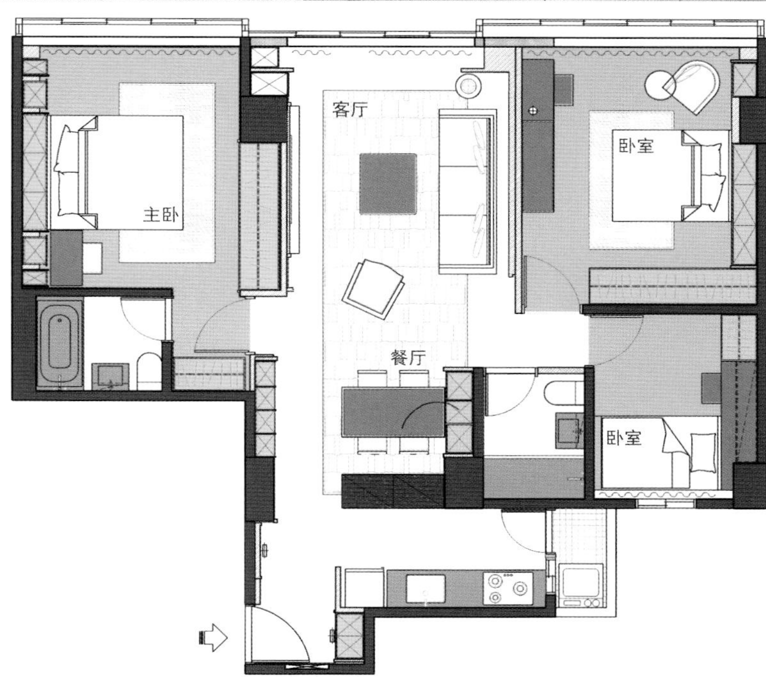

◇ NATURAL LIGHTING 自然采光

In order to solve the problem of low ceiling of the old house, the designers skillfully use the technique of visual extension to widen the whole space, which not only makes the living room field more spacious and magnificent, but also extends the living room functions; sun lights outside the window sunshine flood into the house, bringing visual and psychological warmth and enjoyment.

为解决老屋常见的天花低矮等问题，设计师巧妙利用视觉延伸的手法，拉阔整体空间感。不仅让客厅场域更加开阔大气，而且延伸客厅机能，让窗外阳光能大片泻入屋内，带来视觉与心理上的温暖享受。

迭影 Overlapping Shadows

设计公司：慕泽设计
Design company: MUZE design

项目地点：台湾台北
Location: Taipei, Taiwan

项目面积：120m²
Area: 120m²

主要材料：凿面石、大理石、茶镜、木地板、钢刷木皮、进口薄石材、烤玻等
Main materials: pitch-faced stone, marble, tawny glasses, wood floor, steel brush wood veneer, imported thin stone, burnt glass, etc.

DESIGN CONCEPT 设计理念

The comfort and practicability of modern style have been widely preferred by residents. Muze Design defines design of this project as modern style to make the owner appreciate the warmth of home in nature. Overlapping shadows stagger and the virtual and actual situations complement each other, creating a changing and concluding dramatic effect in the limited space. Stones send out artistic breaths, inject aesthetic implications into decorations of the whole space.

Modern yet not simple, delicate and natural, the space field continues the aura convection at the same time clearly defines public and private areas. Large area of French window has a common border with the outdoors, bringing natural lights and scenery.

现代风格的舒适和实用一直广受居住者的喜爱，慕泽空间将此住宅设计为现代风格，让屋主于自然中领略家的温情。迭影交错，虚实相生，在有限的空间条件限制下有了起承转合的戏剧效果，石材之间互相散发出艺术气息，为整个空间的装饰纳入美学意涵。

现代却不简单，精致不失自然，空间场域延续气场对流，同时界定出公私领域的明确分界。大面积的落地窗和室外接壤，引进纯天然光线和景致的自然幸福感。

◇ SPACE PLANNING 空间规划

Scenery changes at every step. Entering from the porch into interior, you can see the kitchen and dining room. Though it is just a flat residence, the designer reasonably plans the layout to make clear priorities, to separate public and private areas and to avoid chaos of life. The living room, dining room and kitchen are open. What's more, there is a bar in the kitchen to strengthen the sense of participation of family or guest when cooking and to taste wine or eat breakfast here. On one side of the kitchen, there is a master room with bedroom, cloakroom and bathroom , creating the most comfortable living space for the owner. On the other side, there is a small study with a bed which can be upfolded and provides a rest space for the tired owner when reading. The whole space layout makes full use of every functional area.

移步换景，从玄关进入室内，便能看见厨房和餐厅。虽然只是平层住宅，但设计师在空间规划中合理布局，使得主次分明，公私领域错开，避免了生活中交错的混乱场面。客厅、餐厅和厨房呈开放式的格局，厨房还设有吧台，加强烹饪时家人或者客人的参与感，同时还能在此品酒吃早餐。厨房一侧主卧房，卧室、衣帽间、卫浴室一应俱全，为主人打造最舒适的居住空间。室内另一侧则是一个小空间的书房，上掀式的收纳床可给在此阅读疲劳时的主人一个短暂的休息时间，整体空间布局让各个机能区域发挥了最大的功能作用。

◇ NATURAL COLORS 自然色彩

Entering interior, you can see natural and plain colors which are mainly in a cold tone which injects reasonable concise breath for the space. White sunken ceiling and dark brown floor form a contrast, creating a stable atmosphere. The TV background wall made of gray marble and the bar echo with each other and have primitive charms. White leather sofa matches with black velvet pillows, making the contrast more vivid and breaking connections between colors. The dining table and chairs also choose two kinds of colors to make a contrast; one type is black concise chair, the other is transparent and postmodern chair; they form a strong artistic tension and resonance.

进入室内便能感知到自然纯朴的空间色彩，以冷色调为主，为空间注入理智简约的空间气息。白色下沉式的吊顶，和深棕色的地面形成对比，营造出稳重的氛围。灰色大理石的电视背景墙和吧台，相呼应又具有原始的韵味。白色皮革沙发搭配黑色毛绒靠枕，使得对比更加鲜明又打破色彩之间联系。同时餐桌椅也选择两种颜色作为对比，一种是黑色简约式的椅子，另一种是透明具有后现代意味的椅子，两者之间形成强烈的艺术张力和共鸣。

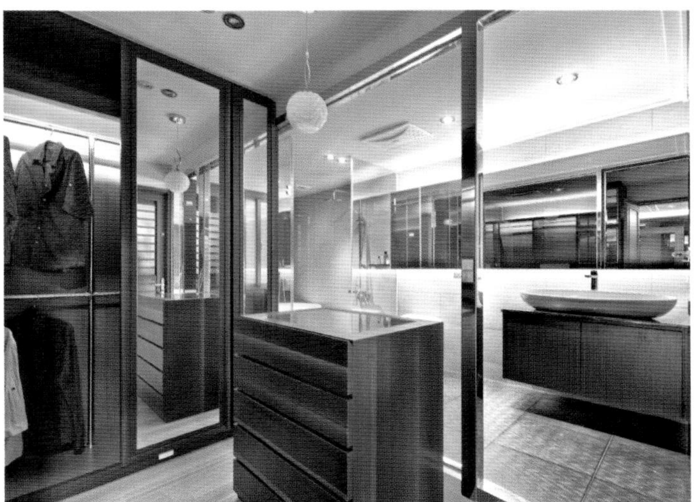

◇ LIGHTING DESIGN 照明设计

Apart from natural lighting, the interior lighting aims to add a warm atmosphere for the family. Whether it is scattered tube light or light luxurious chandelier, they add temperature for the family and light up warmth, in addition to the space. The flickering candle lamps on the dining table and bar and table lamp on the master bedroom bedside create a rich and warm atmosphere for life and make the whole family full of human love.

除去自然的采光，室内照明设计旨在增添更多家庭温暖的氛围。无论是分散开来的筒灯，还是轻奢风格的吊灯，除去照明之外，更多的是为家庭增加温度，点亮温暖。餐桌和吧台上星星摇曳的烛灯，主卧室床头的台灯，又为生活营造富有温情的氛围，让整个家更充满人性的关爱。

森林里的时尚经典

Fashion Classics in the Forest

设计公司：朱英凯室内设计事务所
Design company: KEI Design Studio

设 计 师：朱英凯、吕柏仪
Designers: Kevin Ju,Vivian Lu

项目地点：台湾台中
Location: Taichung, Taiwan

项目面积：175m²
Area: 175m²

主要材料：曙光大理石、珊瑚洞石、进口抛光石英砖、钢刷柚木皮、钢刷白橡木、镀钛铁件、酒精壁炉、特殊漆、皮革裱布、壁纸、超耐磨地板等
Main materials: dawn marble, coral travertine, imported polishing quartz brick, steel brush teak wood veneer, steel brush white oak, titanium iron, alcohol fireplace, special paint, leather cloth, wallpaper, super wear-resistant floor, etc.

摄 影 师：吕国企
Photographer: Guoqi Lv

DESIGN CONCEPT
设计理念

Because of work, the owner has to leave Taipei and Taoyuan to settle in Taichung and chooses this “into the forest” house whose facade looks like a vertical forest. The space planning and material selection are very exquisite. So how to inject luxurious and natural elements into the space for the fashionable owner becomes the most important task.

The TV wall of the living room uses linear cutting to show the bottomless “forest” which is the nonnegligible beautiful scenery in the space. The sofa background wall uses storage functions to create a circular layout, with a non-flammable smoke fireplace embedded in the back which slowly releases the temperature in the coral white marble. Back to the book wall and sitting in front of the fireplace, with the green scenery outside the window and marble like the forest, one can start a tranquil afternoon reading. In the dining and kitchen area which is in the same axis as the reading area and links green with lighting, natural humanistic temperature of wood and iron reconciles the avant-garde stainless steel and stone platform; the dining and cooking functions are divided by materials; what’s more, the place where collects the owner’s mugs hides the only large cylinder of the space.

屋主因工作辗转由台北、桃园来到台中定居，选择这处以“入深林”为题，外观宛如垂直森林的建案为最终落脚处，在空间规划及选材皆十分讲究。因此如何为时尚的屋主，于空间中纳入奢华与自然调和的风格元素，成为最重要的课题。

客厅电视主墙以利落时尚的石材，透过线性切割表现深不见底的“森林”，为空间不容忽略的美丽景致。沙发背墙以收纳创造环状动线，背面嵌入无燃烟的酒精壁炉，于珊瑚白大理石中缓缓释放温度。背对书墙坐在壁炉前，挟伴着窗外绿意与宛如森林的曙光大理石，宁静的阅读午后就此开展。与阅读区位于同一轴线上，串连绿意采光的餐厨空间，木质与铁件的自然人文温度调和前卫的不锈钢、石材餐厨台面，将用餐与煮食的机能以材质划分，并于展示屋主马克杯收藏的机能之中，藏起空间唯一的大柱体。

SHOWROOMS

◇ NATURAL COLORS 自然色彩

In bedroom for the parents, mellow sapphire TV wall becomes a fashionable main vision of the space. In addition to sleeping area, there is a binary dressing and storage space outside for them to store clothes. The daughter's room uses the fashionable and classic Hermes orange; the cubic cutting lines are used in elevations of the bedside and bookshelf.

父母居住的卧房，醇厚的宝蓝色电视墙成为休憩空间的时尚主视觉，除卧眠区外空间规划二进式的更衣收纳，满足两人的衣物收纳量。女儿房以时尚经典的爱玛仕橘为用色，如城堡般的立方切割线条，用于床头、书架等立面。

◇ SPACE PLANNING 空间规划

In addition to the basic configuration of the living room, dining room and kitchen area, in order to correspond to the family of three, the designers change the original pattern of three rooms into double master bedrooms which are almost in symmetric pattern and are equipped with independent dressing spaces, only a slight difference in function and size of the bathroom.

本案除基本配置的客厅、餐厅、餐厨区等空间外，为了对应一家三口单纯的人口结构，将原先三房的格局改置为双主卧空间，几乎对称的空间格局，皆备有独立的更衣空间，仅有卫浴机能与大小的些微差异。

◇ LIGHTING DESIGN　照明设计

"The journey of exploring the forest" starts from the porch. Sights start from the suspended collecting cabinet and three-dimensional sculpture wood elevation, then are guided by skylight bands to settle in the lighting scenery. Coming to the living room across the porch, the unique texture and straight cutting concave and convex surface of dawn marble is like a bottomless forest under the foil of natural lights and indirect lights above.

"探索森林之旅"由玄关开始，悬浮的收纳量体与立体雕塑的木质立面，循着天际光带的引导，将视线落定在采光景致中。越过玄关来到客厅，曙光大理石的独特纹理、直向切割的凹凸面，在自然光与上方间接灯照射下，仿佛一座深不见底的森林。

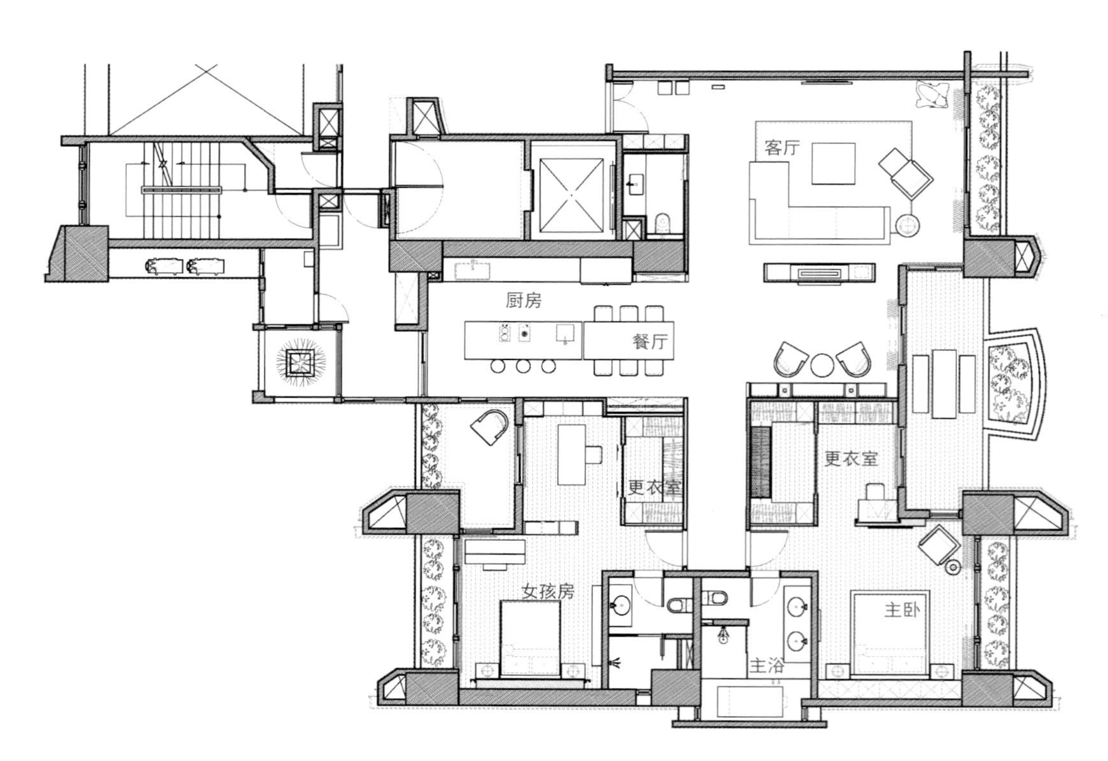

凌静 Tranquility

设计公司：慕泽设计
Design company: MUZE design

项目地点：台湾台北
Location: Taipei, Taiwan

项目面积：120m²
Area: 120m²

主要材料：大理石、木皮、仿古镜、喷漆等
Main materials: marble, wood veneer, antique mirror, spray lacquer, etc.

DESIGN CONCEPT 设计理念

The rugged wooden ceiling is as continuous as mountains, starting from the porch to the living room, dining room and kitchen, which connects a vivid vision of the whole field. The antique mirror near the main TV wall symbolizes the lake imagery, making the space like a landscape painting and showing elegant and powerful momentum of the field. This is an renovation project for a 15-year-old house. Inspired by natural landscapes, MUZE design team uses line modelings and material changes to hint images such as mountain, soil, sunshine and lake, making the house get rid of past limits and depressive sense of space to thoroughly show the luster and vitality.

高低起伏的木质天花造型，从玄关开始游走至客厅、餐厅及厨房，有如山峦般绵延不断，串联起整个场域的生动视觉。电视主墙旁的仿古镜则象征湖泊意象，让空间有如一幅山水画，展现优雅大器的场域气势。这间是已有十五年屋龄的老屋翻新案例，慕泽设计团队以自然界的天地景观为设计灵感，透过线条造型与材质变化，暗喻山峦、土壤、阳光及湖泊等万物意象，让老屋挥别过去限制式、压迫的空间感，彻底改头换面展现光采与活力。

◇ SPACE PLANNING 空间规划

Entering from the porch, the living room, dining room and kitchen adopt open design to make the space more capacious and use floor materials to divide the field. Upon the dining table, there is a round glass droplight whose smooth and glittering texture brings light and transparent effects, making the finishing point of the space vision. In the public area, the broken line elements continue to spread to the bedroom, which makes the theme design elements echo with each other. The bedside background in the master bedroom is covered with marble and plated titanium metal, creating an elegant and magnificent exquisite texture. The design of the secret door is made with ulterior motives, which seems to be a complete antique mirror elevation, but hides an entrance to the gym and bathroom. The secret door design makes the layout more upright and foursquare and shapes a harmonious visual aesthetics.

从玄关进入室内，客厅、餐厅与厨房采用开放式设计，让空间感更加宽敞，并借由地坪材质作为场域划分。餐桌上方选搭圆形玻璃吊灯，其光滑晶莹质感，带来轻盈透亮的效果，为空间视觉画龙点睛。公领域的折线造型元素，持续蔓延至卧室内，让主题设计元素相互呼应。主卧室床头墙采用大理石与镀钛金属做拼接，创造优雅且大器的细腻质感。暗门设计别有用心，看似完整的仿古镜立面，实则隐藏通往运动室与卫浴的入口，借由暗门设计让格局更加方正，且形塑和谐一致的视觉美感。

◇ DECORATIVE MATERIALS　装饰材料

The whole ground is covered with wood floors, symbolizing the soil that gives birth to life and bringing people soothing and warm power. The sofa background is decorated with antique brass surface, representing the warm sunshine; some parts collocate with titanium metal closing, adding texture to the whole elevation design. A whole piece of antique mirror is put in the wall facing the marble TV background wall, reflecting the interior furnishings by its reflective characteristics, which is like a reflection on the lake, adding a hazy aesthetic feeling to the space.

全室地坪采木地板铺陈，象征孕育生命的土壤，带来安定人心的温润力量；沙发背墙则饰以古铜质感表面，代表温暖的阳光，局部搭配镀钛金属收边，让整体立面设计更添质感；面向大理石电视主墙衔接一整面仿古镜，借由其反射特性照映出室内陈设，有如湖面上的倒影，为空间覆上一抹朦胧美感。

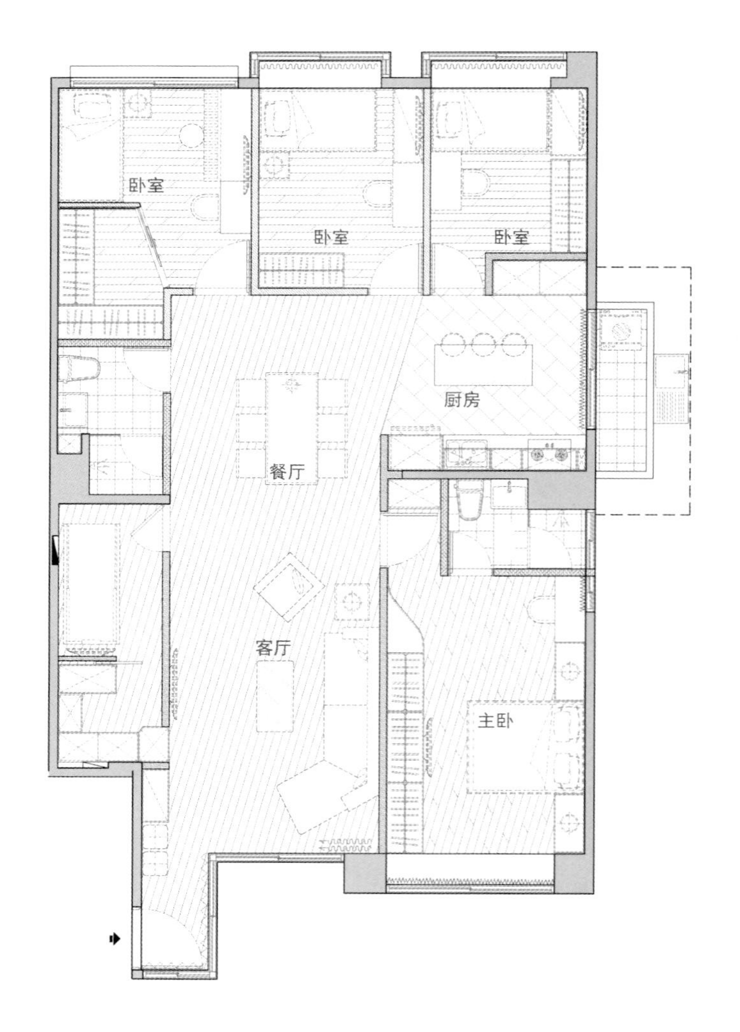
卧室
卧室
卧室
厨房
餐厅
客厅
主卧

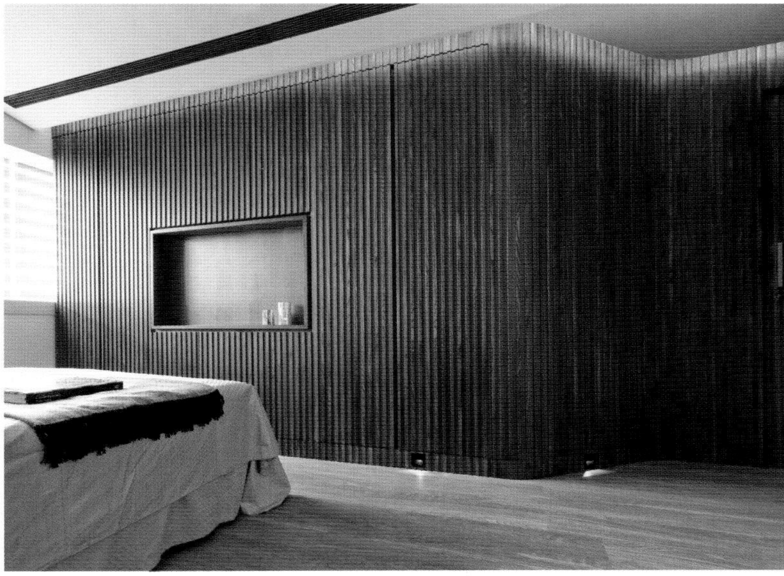

◇ NATURAL LIGHTING 自然采光

Because the old house has many beam structures, MUZE design team especially uses rolling broken line modeling to decorate the ceiling and beam lines, maximizes the house height, avoids the oppressive feeling caused by low beams and makes lights flow freely in the space. Except for bringing plentiful lights by large areas of window, MUZE design team indirectly embeds lights on both sides of the ceiling, imitating the refraction effect of sunshine and making the space time dynamic and vigorous.

因老屋原有许多梁柱结构，慕泽设计团队特别以起伏折线造型修饰天花与梁线，将屋高视觉拉至最高，避免矮梁造成的压迫感，同时也让光线在空间自由流动。除了大面开窗引入丰沛采光外，慕泽设计团队亦在天花两侧嵌入间接光源，仿照出阳光折射效果，让空间时刻充满活力生气。

感性·法国女仕宅邸

Sensibility, French Lady Mansion

设计公司：缤纷设计
Design company: L'atelier Fantasia

设 计 师：江欣宜
Designer: Idan Chiang

项目地点：台湾台北
Location: Taipei, Taiwan

主要材料：灯饰、布艺、镜面材质等
Main materials: lighting, cloth, mirror, etc.

DESIGN CONCEPT
设计理念

Based on the latest completed Ritz Carlton hotel in Paris, France, the designer uses a lot of light blue line boards, pillows and cloth to present the exquisiteness and nobility like the French hostess and the Versailles Palace. The attention to details and crafts is shown in Italian and European furniture, lighting and cloth. The space is decorated with Hermes scarves which are collected by the hostess and framed into art works, presenting the requirements and extension of crafts incisively and vividly. The space integrates furnishing details such as color, cloth, lamps and lanterns, flower art and plants, creating an elegant living space and a luxurious and retro mansion.

以法国巴黎 Ritz Carlton 最新装修完成的饭店为基调，大量运用淡蓝色系的线板、抱枕、布艺，串联出如同法国女主人和凡尔赛宫廷的精致与高贵。对于细节及工艺的重视，展现于使用意大利及欧洲的家具、灯饰及布艺。空间中点缀着女主人收藏的爱马仕丝巾，将其裱框成艺术品，将工艺的要求及延伸发挥的淋漓尽致。空间汇整颜色、布艺、灯具、花艺与植栽等陈设细节，打造优雅的生活空间与奢华复古的豪邸风采。

◇ SPACE PLANNING 空间规划

Entering into the room which is full of French amorous feelings, the porch welcomes you with elegant pine stone green curtains and long sofa. The living room and the study are connected in an open form, symbolically separating the flexible spaces. The black fireplace is exquisite and mysterious; the sofa in the living room is given priority to the blue in dark and light gradation, dignified and fashionable. The dining room is close to the living room and separated by the door; dead branch droplight is full of artistic design feeling. The master bedroom and the secondary bedroom are simple and comfortable, giving off a tranquil fragrance.

进入充满法式风情的室内，玄关以高雅的松石绿窗帘和长条沙发凳相迎。客厅和书房以开放形式相连，灵动的空间象征性地分隔。黑色壁炉精致神秘，客厅沙发以深浅渐层的蓝色为主，端庄不失时尚。餐厅以门相隔紧挨客厅，枯枝吊灯富有艺术设计感。主卧和次卧则简单又舒适，散发出静谧的幽香。

DIOR NEW COUTURE

◇ DECORATIVE MATERIALS AND NATURAL COLORS 装饰材料、自然色彩

In the modern and luxurious living room, sofa and chairs use dark and light blue velvet, maintaining the color consistency; pillows in layered effect also become the focus of the whole space. Curtains of the porch and chairs use the same kind of cloth, showing luxuriant texture and making people feel as if in the world of upper class ladies.

In the dining area, the dining chairs use Chanel knitting structure texture fabric with jumping color rolloff in the back and collocate with European unique branch lighting, echoing with flower art and green plants. The planning of colors and details connects the consistency of the whole space, making the space comfortable without oppression and adding rich saturation and energy into the mansion. In order to meet the visual enjoyment and functional demand, the study and the living room are planed in the half-open pattern and use four mirror sliding doors as the partition to enlarge the vision of them.

在摩登、奢华的客厅区域，以深浅渐层蓝色绒布为主沙发及单椅的材质，延续色彩的一致性，渐层效果的抱枕，亦成为整体空间的聚焦点。玄关窗帘与玄关椅使用同款布料，展现出华丽的质感，让一进门的视觉感官体验宛如进入上流女士的世界。

餐厅区餐椅设计以香奈儿粗针织构质感为主的布料，背椅滚边的跳色处理，搭配上欧洲工艺独特的树枝状灯饰，与花艺与绿色植栽相互呼应。透过颜色及细节的规划，串联起整体空间的一致性，使空间舒适无压迫感，宅邸添加丰富的彩度与能量。满足视觉飨宴与机能需求，书房与客厅以半开放式的规划结合，运用镜面材质的四扇推拉门作为两者间的芥蒂，放大书房与客厅的空间视角。

典藏·品味 Collection and Taste

设计公司：成晟室内装修设计工程有限公司
Design company: CHENG SHENG INTERIOR DESIGN

设 计 师：李兆亨、成秋双
Designers: Li Tsan Hen, Cheng Chiu Shuang

项目地点：台湾台北
Location: Taipei, Taiwan

项目面积：234m²
Area: 234m²

主要材料：大理石、踱钛、线板、KD木皮、仿古镜等
Main materials: marble, titanium, line board, KD wood veneer, antique mirror, etc.

摄 影 师：郭家和
Photographer: Ar-Her Kuo

DESIGN CONCEPT 设计理念

Open the door, with the transfer of vision, you can enter into a feast of beauty. Cheng Sheng Interior Design uses “low-key luxury” to present texture of the space, follows the couple’s life habits and style preferences and their aesthetics formed by living in a foreign country, plans the space grain which conforms to the human nature through detailed analysis and integration and creates a comfortable and beautiful environment to make the couple gracefully immerse in the space.

The designers use the application of materials and the pavement of style to simplify complicated elements of neo-classical style and use simple line board modeling wall to lead out neo-classical atmosphere in the field. For example, in the living room, the whole piece of marble TV wall is used to add luxurious sense and charm and to create an exalted and high texture life. The bedroom uses hotel texture to create sleeping style; every furnishing reveals the elegant and noble temperament in it.

推开门扉，随着视线的移转，我们进入了一场关于美的盛宴，成晟设计以“低调奢华”铺述空间的质感，并依照男女屋主的生活习惯及风格喜好，以及长期旅居国外的美感薰习，透过点点滴滴的剖析与整合，规整出符合人性的空间脉络，营造舒适与美感兼容的环境，让屋主两人能够更优雅地沉浸在空间之中。

设计师借由材质的运用及风格的铺陈，简化新古典的繁复元素，使用简约线板的造型壁面，引出场域中的新古典氛围。如在客厅用整面式的大理石电视墙来增添奢华感与气韵，缔造尊贵高质感的生活。而在卧室则以饭店的质感打造寝眠风格，一物一饰，均彰显了居室蕴藏于内的高雅贵气。

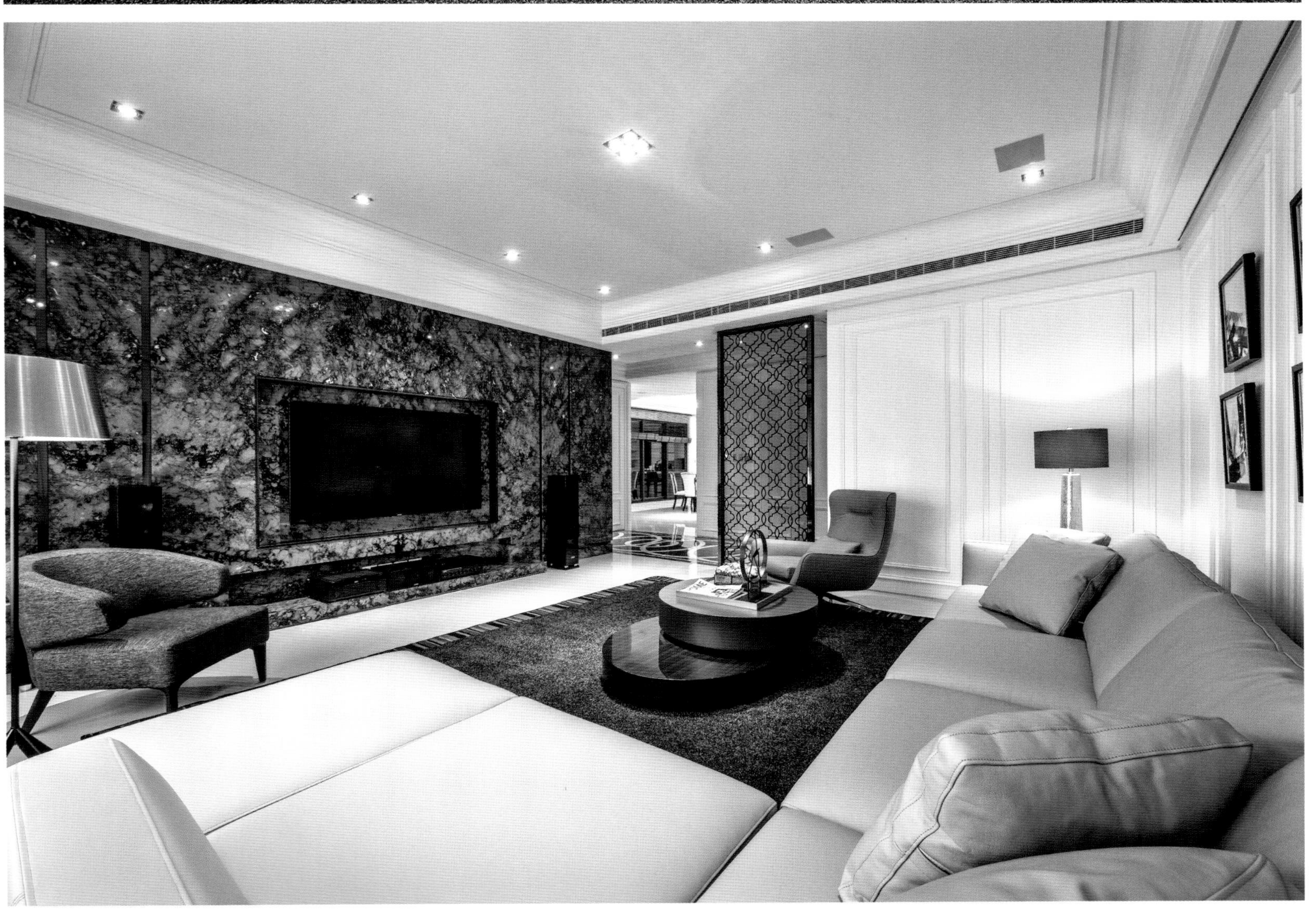

◇ DECORATIVE MATERIALS 装饰材料

The welcome image after entering the door develops the design prelude in a beautiful appearance. Seamless marble parquet floor matches the side view constituted of mirror, iron, stone and crafts, in addition to the ceiling and walls piled by line boards, perfectly integrating classics with art and forming a contemporary masterpiece.

入门后的迎宾意象，以绝美之姿展开设计序曲，无缝水刀大理石拼花的地坪，搭配由镜面、铁件、石材与艺品构织的端景，加上线板堆栈的天花与壁面，完美融合经典与艺术，形塑一幅集当代的巨作。

◇ SPACE PLANNING 空间规划

In order to conform to the layout, the designers set the large-scale living room and dining room in left and right side of the porch respectively; the pattern of functional partitions keeps the completeness of utilization; the circulation and vision penetrate without mutual interference, but with infinite and extensive spaciousness. At the same time, the bar and light food area combine in an open form, enlarging the visual depth of field and extending the space scale.

为了顺应动线，设计师将大尺度的客厅及餐厅，分别建构在玄关的左右两侧，机能分区的格局配置，在使用上保持完整性外，动线和视觉穿透而互不干扰，却有无限延伸之宽广性。同时以开放形式结合后方吧台与轻食区，除了放大视觉景深外，也延伸了空间尺度。

◇ NATURAL LIGHTING 自然采光

The designers make full use of lights from the balcony to bring outdoor warm sunshine through the French window into the study and dining bar area, adding happy temperature to the field.

设计师充分利用露台的光源，将户外暖阳透过落地窗援引入内，围绕在书房与餐吧区，增添场域的幸福温度。